# Synthesis Lectures on Engineering, Science, and Technology

The focus of this series is general topics, and applications about, and for, engineers and scientists on a wide array of applications, methods and advances. Most titles cover subjects such as professional development, education, and study skills, as well as basic introductory undergraduate material and other topics appropriate for a broader and less technical audience.

John Krupczak, Jr.

# Understanding
# Technological Systems

 Springer

John Krupczak, Jr.
Department of Engineering
Hope College
Holland, MI, USA

ISSN 2690-0300          ISSN 2690-0327  (electronic)
Synthesis Lectures on Engineering, Science, and Technology
ISBN 978-3-031-45440-0          ISBN 978-3-031-45441-7  (eBook)
https://doi.org/10.1007/978-3-031-45441-7

This Springer imprint is published by the registered company Springer Nature Switzerland AG
The registered company address is: Gewerbestrasse 11, 6330 Cham, Switzerland

Paper in this product is recyclable.

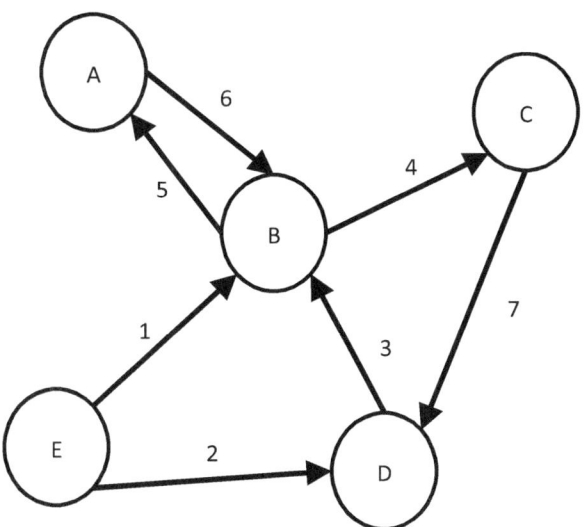

*For my parents*
*John and Dolores Krupczak*

# Preface

*It seems every mature science has been able to identify its basic elements (atoms, quarks, genes, etc.) and to explain its phenomenon as the known interactions of these elements.*

*Bela Julesz*

*Therefore, the (engineer) should sit down among levers, screws, wedges, wheels, etc. like a poet among the letters of the alphabet, considering them as the exhibition of his thoughts; in which a new arrangement transmits a new idea to the world.*

*Robert Fulton, A Treatise on the Improvement of Canal Navigation (1796)*

## What Is This Book About?

This book is about understanding technology using the perspective of systems. A definition of a system is a group of interacting entities, forming a network, and achieving a common outcome. Here, the interacting elements are identified as the "components" of a technological system. A technological system is considered to be any human-made physical object intended to achieve a particular purpose. Technological systems encountered in daily life range in complexity from a hairdryer to a mobile phone. A component in this instance means a physical object created to provide a specific function such as a light-emitting diode, a battery, a beam, or an internal combustion engine. This framework views technological systems as created using components to provide specific capabilities or functions. Components contributing well-defined functions interact with other components to create systems offering value or usefulness that exceeds that of individual components considered individually. Technological systems evolve to create clusters of related systems meeting unique applications through variations on a core set of components and system arrangements.

This work is intended to fill the gap between engineering science and engineering design. Engineering science with various branches and sub-fields offers precise mathematical analyses of particular circumstances and specific systems. Engineering design provides methodologies to identify requirements and efficiently create designs that meet specifications. Understanding technological systems offers a consistent framework for developing a working knowledge and vocabulary of actual real physical components and currently existing systems. These existing components and systems form the building blocks of new designs and analyses.

The prospect of understanding technology is challenged by the immense variety of human-created objects such as personal electronics, medical devices, automobiles, civil infrastructure, and spacecraft. A search for some element of commonality identifies that each of these can be characterized as a network of interacting components working together to achieve an overall function. Applying a system perspective to describing and understanding technological objects appears reasonable given that the field of biology follows a similar approach to describing and analyzing the diversity seen in ecosystems of living organisms.

Understanding technology using the perspective of systems offers an advantage of not requiring extensive prerequisite knowledge of mathematics or specific engineering science fields. The approach taken here is intended to be accessible to anyone seeking to be better informed about technology and is not written just for those specializing in engineering and technological fields. The material aims to provide a step-by-step explanation of one type of organizational structure for technological systems and to develop foundational knowledge that readers can later build on through their own interests and experiences with our pervasive technology.

## Overview of Topics

Chapter 1 *Technology: Form, Function, Value* describes background concepts regarding the view of human-made technology as changes in the physical world intended to accomplish a specific purpose. The concepts of form and function are introduced as a perspective through which to interpret the physical world. Presently most technology is not made by the end user, and the issue of requirements and specifications is reviewed. An emphasis is made on the difference between requirements for the function of the system, what the system must do, versus requirements on the form of the system, or its physical characteristics.

Chapter 2 *Systems Transforming Materials, Energy, and Information* introduces the concept of a system and describes how the general features of a system can be applied to technological objects. Systems are distinguished from their external environment by a system boundary. Systems transform available inputs into desired outputs. For technological systems, the inputs and outputs can be broadly classified as energy, materials, and information.

Chapter 3 *How it Works: Components and Subfunctions* explains technological systems as consisting of components that provide subfunctions inside the system boundary and interact to transform input into outputs. An individual component carries out some subset of the transformations needed to produce the overall functions of the system. The idea of describing "how something works" is developed as identifying the major components and their interactions necessary to accomplish overall system functions. A process is described for constructing a diagram of a system of components and interactions illustrating how inputs are transformed into outputs.

Chapter 4 *Phenomena and Models* focuses on the component as an embodiment of physical phenomena. Components utilize natural phenomena in carrying out functions within technological systems. The component's physical characteristics, or its "form," are determined by the physical processes employed by the component in accomplishing its functions within a system. Mathematical models can be developed to characterize component and system behavior.

Chapter 5 *System Characteristics in a Technological System Context* demonstrates how features considered characteristics of systems are observed in technological systems. Distinctive features of systems include the concepts of emergence, hierarchies, dynamic behaviors, feedback, homeostasis, and nonlinearity. Also introduced are topics of component autonomy and system failure analysis.

Chapter 6 *Component Parameterization and Transfer* examines how components transfer across systems. Modern technological systems use the same component to carry out the same function in different technological systems. The chapter describes how components are parameterized to achieve the same function in systems with different requirements. Standardization of components and manufacturer's data sheets are explained. The idea of a "generic data sheet" is introduced as an outline for developing a working vocabulary of technological components.

Chapter 7 *System Interdependence* acknowledges that technological systems do not operate as isolated entities but depend on other technological and non-technological systems. Systems feed, make, and govern other systems. Because technological systems are actual physical objects, the nature of the physical world is changed when new technology is created leading to possible changes in human behavior. Areas are noted where human agency has potential opportunities to influence the nature of technological systems.

Chapter 8 *System-Level Similarity: Technological Domains or Families* looks at how groups of technological systems develop around a core system and the application of the same underlying physical principles. Systems within a domain address related problems by modifying a basic system through means such as changes in scale, addition of functions through specific components or subsystems, and adjustment of component forms. Industries and engineering subfields often develop around technological domains. The chapter introduces codes and standards as a means of ensuring safety and efficiency for system performance within a domain.

Chapter 9 *Technological Evolution and Innovation* describes some features of how technology changes from a component-system perspective. Technological system evolution can frequently be described as modification at the component level. Changes affecting multiple components commonly involve changes in the underlying principles used to carry out a system subfunction. Patents are introduced as representations of technological components and systems deemed novel compared to the existing state of the art.

## Background on the Development of This Book

The materials in this book were initially created for a course on technology for liberal arts students that I developed at Hope College. In first preparing the class, I could not identify any existing organizing principles to use to structure the course that were not just diluted engineering science topics or simply a collection of various unrelated technologies. At the same time, while conducting courses in engineering design and introduction to engineering for engineering students, I experienced the challenge of developing engineering students' familiarity with actual existing components and systems which they could use in carrying out their own design activities. The approach used here for understanding technology as a system of interacting components was developed over two decades of gradual progress to help both liberal arts and engineering students to achieve a more empowered relationship with the actual technologies encountered in everyday life. The materials here incorporate concepts from the fields of product development, design theory, systems engineering, and the engineering sciences modified to help explain how existing technologies combine disparate elements and phenomena to meet human wants and needs.

Holland, MI, USA                                                          John Krupczak, Jr.

# Acknowledgements

This project was many years in development, and I would like to thank the many individuals who contributed to the final outcome as presented here. I am particularly indebted to John Heywood and Mani Mina for their ongoing encouragement and assistance. In addition, David Ollis provided vital guidance in the initial phases of this work.

Many people at the American Society for Engineering Education, and particularly the Technological and Engineering Literacy/Philosophy of Engineering Division, contributed productive discussions and feedback on ideas regarding technological systems, technological literacy, engineering education, and the philosophy of engineering. This includes John Blake, W. Bernard Carlson, Alan Cheville, Soheil Fatehiboroujeni, Steve Frezza, Jerry Gravender, Katherine Goodman, Joseph Herkert, Carl Hilgarth, J. Douglass Klein, Russ Korte, Kathryn Neeley, John R. Reisel, Karl Smith, and Julia Williams.

Some topics presented here were identified and refined during the course of research projects developing and advancing technological and engineering literacy. I am grateful to my collaborators in these efforts and would especially like to thank Matt R. Bohm, Kate Disney, Yuetong Lin, Julie S. Linsey, Alexander R. Murphy, Robert L. Nagel, A. Mehran Shahhosseini, Scott Vanderstoep, and Leslie Wessman.

A number of student researchers made contributions in creating, testing, and improving educational materials and projects to advance approaches to explaining technological components and systems. I am appreciative of the assistance from Lauren Aprill, Ovais Aulakh, LaToya Austin, Tucker Babb, Nathan Bair, Tim Benson, Nathan Ceja, Dale Corleu, William Day, Matthew Dickerson, Adriene Elinski, Christian Forester, Luke Hoogeveen, Kaitlyn Kopke, Casey Lamb, Dan Langholtz, Kristen Lantz, Joel Lanus, Daniel Lappenga, Nathaniel Makowski, Ben Maninno, Dave Muir, Catherine Otto, Leah Patenge, Julian Payne, Luke Pinkerton, Mary Sheryl Ramesh, Daniel Rodak, Stephanie Ross, Courtney St. Clair, Elisabeth Salazar, Matthew Scholtens, Ashely Shaneck, Dale Shepherd, Alexander Sherstov, Aaron Silver, Danielle Simmons, Marissa Steffens, Samantha Steffens, Olive Stohlman, Matthew Stolz, and Kristi Van Dyk. Hope College students and Mani Mina's students at Iowa State University provided feedback and suggestions about topics and effective approaches for conveying aspects of technological systems to broad audiences.

In the completion of this effort, I am indebted to the ongoing support and suggestions throughout the process from Dave Amlicke, Dave Brown, Joe Houley, Gul Kremer, Vidu Kulkarni, Gary Lichtenstein, Michael Roppo, Donglu Shi, Carol Sliwa, Keith Sverdrup, Paul Tymann as well as my colleagues at Hope College.

I would like to thank my family for their ongoing support. My wife Catherine Brooks helped with the preparation of the manuscript, Alice Krupczak assisted with photography, and Emmett Krupczak provided comments on the draft.

Some materials in this book are based on work supported by the National Science Foundation under award numbers 9752693, 0341998, 0633277, 0920164, 1121464, and 1650889. Any opinions, findings, and conclusions or recommendations expressed in this material are those of the author and do not necessarily reflect the views of the National Science Foundation.

# Contents

# List of Figures

# Technology: Form, Function, Value

**1**

## 1.1 Chapter Overview

- Technology changes the physical conditions in which people live. Technology is created for a specific purpose or utility, causing changes in the physical environment.
- The concepts of form and function offer a perspective for interpreting physical objects.
- Form describes the material properties and physical characteristics of an object.
- Function refers to what an object does or can do.
- Form is independent of any particular use or application of the object. Some aspects of form can be modified or altered.
- Form properties such as shape or color facilitate recognition of specific objects. Function describes the role, utility, task, or service that the object can fulfill in a situation or circumstance.
- Function describes what a particular object does or can do in interaction with other objects. Function, what a particular object can do, is determined by the form or the physical properties and characteristics of that object.
- Function is initially an abstract idea given physical implementation through form. A particular form can have multiple functions.
- A specific function can be accomplished by multiple forms.
- Verb-noun pairs can be used to express function.
- Technology is often characterized as existing to solve problems. The function of a technological object can be considered as describing the nature of the problem solved.

Technology is typically not created by the end users of that technology. Technology is developed to meet the needs of the end customers as characterized through design requirements. Requirements that describe what the technology is expected to do address function. Other types of requirements impose expectations on physical characteristics or form.

J. Krupczak, Jr., *Understanding Technological Systems*, Synthesis Lectures
on Engineering, Science, and Technology, https://doi.org/10.1007/978-3-031-45441-7_1

## 1.2    Technology Changing Physical Reality

Technology is the intentional modification and manipulation of the natural world carried out by people to achieve a specific purpose. More nuanced definitions can be given for technology but humans' deliberate alterations of the existing physical environment for practical utilitarian ends can serve as a broad working description of technology. In the modern era, these manipulations and modifications can be extremely complex and involve numerous coordinated activities of humans and their organizations.

The development of technology is a central focus of the engineering profession, although many elements of our socio-economic system are involved in bringing modern technological products to the end-users. The engineering involved in developing technology is sometimes characterized as creative problem-solving. In the context of technology, the problem solutions developed involve actual physical objects of some type.

How can things solve problems? What are some examples of things solving problems and meeting needs? For example, the airplane solves the problem of rapid long-distance travel. The mobile phone provides for the need for individual personal communication and entertainment. An electric utility power plant provides electrical energy to solve the problem of powering electrical devices and equipment. An artificial hip solves the problem of an impaired functioning hip.

The term problem is used broadly to refer not only to actual problems but also to the needs, wants, and desires of people. Not all people's needs and wants can be addressed by the creation of physical objects. However, in seeking a solution to problems technological objects are created that have some purpose or utility related to the resolution of a problem.

Developed as the solution to problems, technology creates new physical reality. Technological systems are actual tangible objects that modify or change the content and circumstances of our physical existence and surroundings. Physical reality is changed in at least two ways. First, the problem has been solved at least temporarily. So, the physical reality that included that problem is now changed to a different state. Second, since technology involves actual real tangible objects, the physical world is different. It now contains a new object of some sort.

It is worth noting that the changed reality including this new problem-solving technology can create impacts and problems that did not previously exist. The mobile phone is a recent example of this phenomenon. This technology adopted by end-users for communication and entertainment results in impacts of tracking and targeted marketing.

Creating technological systems refers to bringing into being something that did not previously exist. It is true that the actual raw materials existed in some form before being utilized in a particular technological device. The engineering and manufacturing process results in a new configuration of the materials in a way that was previously unknown.

Technology is concerned with actual physical objects. However, not every aspect of developing technology involves a physical object. Some aspects of creating and developing actual physical systems involve non-physical elements. For example, the computer code that controls the operation of a device might be considered non-physical. Analysis and calculations used to develop a component may be viewed as abstract and non-physical. However, the ultimate endpoint of technological development is some type of actual physical system that people can see and touch and accomplishes some real-world task whether this is a smartphone, a highway bridge, or a heart pacemaker.

## 1.3 The Form and Function Perspectives on the Physical World

Technology is concerned with actual physical objects. These physical objects involve a diverse range of both naturally occurring and human-made objects. Naturally occurring objects are materials like stone, sand, and wood. Whereas human-made objects compose a gamut of processed and fabricated items such as steel bridge beams, artificial heart valves, and the electronic components in a mobile phone.

One approach to viewing this diverse technological world is through the perspectives of form and function. Form and function represent two different ways to view the same object. A physical object can be interpreted from these distinct perspectives of form and function. Form can be defined as what an object is or the physical characteristics of the object. Function is what an object does or can do in interaction with other objects.

As a start, consider a simple naturally occurring object such as the stone shown in Fig. 1.1. The form of the stone is described by the physical properties of the rock. This includes characteristics such as the mass of the rock, its physical dimensions and shape, surface texture, color, the chemical composition of the constituent minerals, and properties such as the density, heat capacity, and fracture strength. Specific data describe the form of this stone, such as it is circular with a diameter of 100 mm, grey in color, smooth surface, with a mass of 2 kg. Form properties are independent of any particular use or application of the object.

Function describes what the stone does or can do in interaction with other objects. The same rock can be used for various functions. For example, it could be used as a hammer, a construction element in a building, a paving stone in a road, to hold open a door or hold down a stack of papers or to break open nuts when preparing food.

As a second example, consider the I-beam shown in Fig. 1.2. Unlike the stone, the I-beam is not a naturally occurring form but was created by people. The name I-beam is derived from the characteristic I-shape of the cross-section. Description of the form of the I-beam would include properties such as the shape, weight, and dimensions of the I-beam, the material, in this case, steel, from which the beam is made along with the properties of that material. Aspects of the form might also include features such as cost and availability of the material.

The I-beam can be used as a structural element in a bridge. Figure 1.3 shows a highway overpass. In the overpass, I-beams oriented horizontally support the weight of the roadbed and the traffic. In the bridge, the horizontal beams run parallel to the roadbed and traffic.

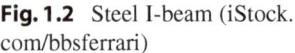

FORM
Size
Weight
Shape
Composition
Color
Specific heat

FUNCTION
Hammer
Support
Construction
Element
Paper weight
Paving stone
Weapon
Food preparation

**Fig. 1.1** Form and function perspective regarding a naturally occurring stone (iStock.com/malera paso)

**Fig. 1.2** Steel I-beam (iStock. com/bbsferrari)

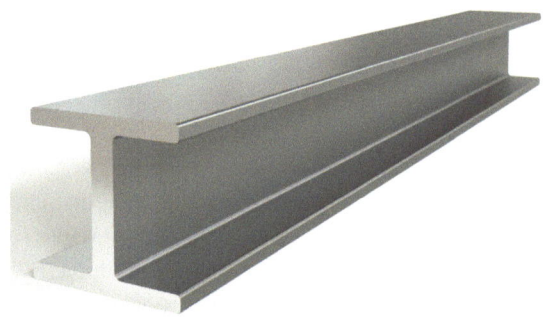

Rather than being situated horizontally as a bridge, the I-beam can be turned on end to support a roof. A vertical I-beam can be used as a roof support rather than a bridge. The I-beam can also serve as the components of a building structure as seen in Fig. 1.4.

## 1.4    Form and Function Multiplicity

Any given form can have a nearly infinite variety of uses. The same object can fulfill multiple functions. The same stone could be used as a hammer, part of a wall, or a door stop. The I-beam can be used horizontally to support a bridge, or vertically to hold up a roof.

Even if an object has a particular intended use, its form properties may enable it to be used for purposes other than its original purpose. As an example, consider a simple

**Fig. 1.3**   I-beam supporting loads in an overpass bridge (iStock.com/taka4332)

**Fig. 1.4**   I-beams acting as columns to support vertical load (iStock.com/Tonylaniro)

manufactured object such as a common GEM paperclip as shown in Fig. 1.5. The form of the paperclip includes characteristics such as its shape, the composition of the steel from which it is made, the diameter and length of the wire, and the mass of the paperclip. The intended function of the paperclip is to temporarily bind together several sheets of paper in an easily applied and removed manner without damaging the paper. It solves the problem of holding sheets of paper together.

The same form of the paperclip can serve many different functions. The form properties of the paperclip lead to it being suitable for other purposes. For example, the paperclip,

**Fig. 1.5** Paper clip used to
hold paper (iStock.com/art
isteer)

in addition to holding paper, can solve the problem of holding a group of keys. The
paperclip could be used as a hanger for small objects. It could be used as a bookmark.
The paperclip can replace a missing tab on a zipper. It could be used to turn a screw.
Even if an object was created for a specific purpose, aspects of the form of the object
may enable it to be suitable for other functions.

The same function can be achieved using different forms. Consider the problem of
eating a salad as shown in Fig. 1.6. A fork is the usual object used to eat salads. However,
the function of eating salad can be accomplished using forms other than the usual fork.
If no forks were available, how could you eat the salad? The salad could be eaten with a
spoon, chopsticks, or tongs. The salad could be picked up with the fingers or liquefied in
a blender and consumed in liquid form. Many different forms can fulfill the function of
eating a salad.

The use of different forms to accomplish a particular function is not confined to solving
the problems of everyday life like salad eating. The same holds true in technological
applications. As another example of different forms that can solve the same problem
consider the problem of supporting a load such as a roof. The same load-bearing function
can be accomplished using a variety of possible forms. An I-beam acting as a column
can provide the function of supporting a load. A stone or concrete column could support
the load. These various forms are possible candidates to solve the problem of providing
support. Figure 1.7 illustrates different approaches to supporting a roof.

In creating technological systems different forms might be used to provide the same
function in different circumstances. This can happen even if one commonly used solution
exists. In technological systems, a frequently needed function is to conduct electric current between two points. Copper wire typically provides this function. However, electric
current can also travel through other metals in various forms. Flat copper strips rather than
round wire are used to conduct electric current along a circuit board. Solder (a mixture of

**Fig. 1.6**  Salad implements **a** fork (iStock.com/tessarola); **b** chopsticks (iStock.com/TayaCho); **c** tongs (iStock.com/Lena_Zajchikova)

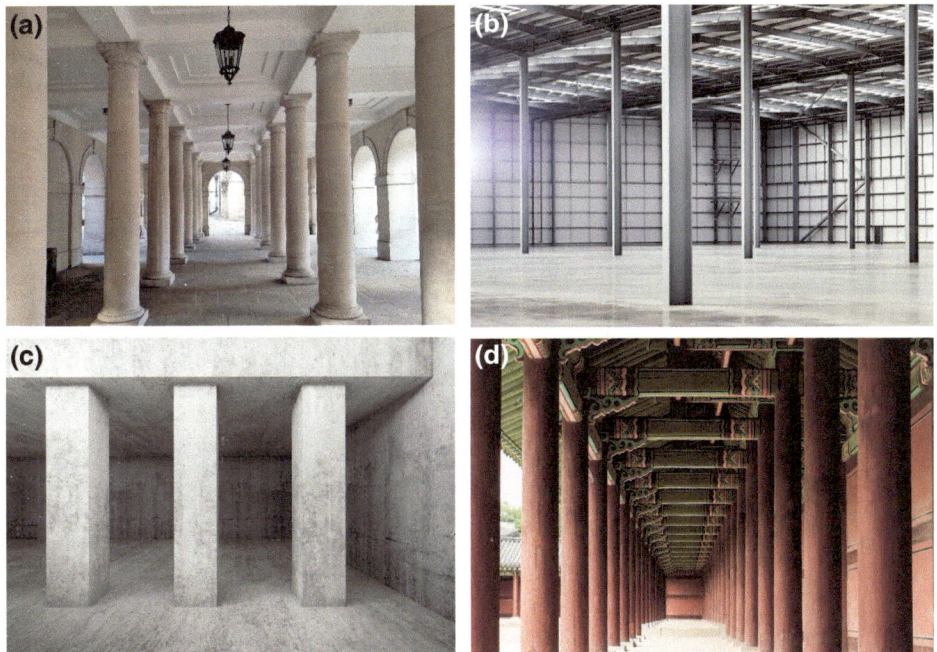

**Fig. 1.7**  Stone, steel, concrete and wood columns supporting vertical loads: **a** marble columns (iStock.com/araelf); **b** steel I-beams (iStock.com/Daniel Graves); **c** concrete columns (iStock. com/Evgeny Sergeev); **d** wood columns (iStock.com/artapornp)

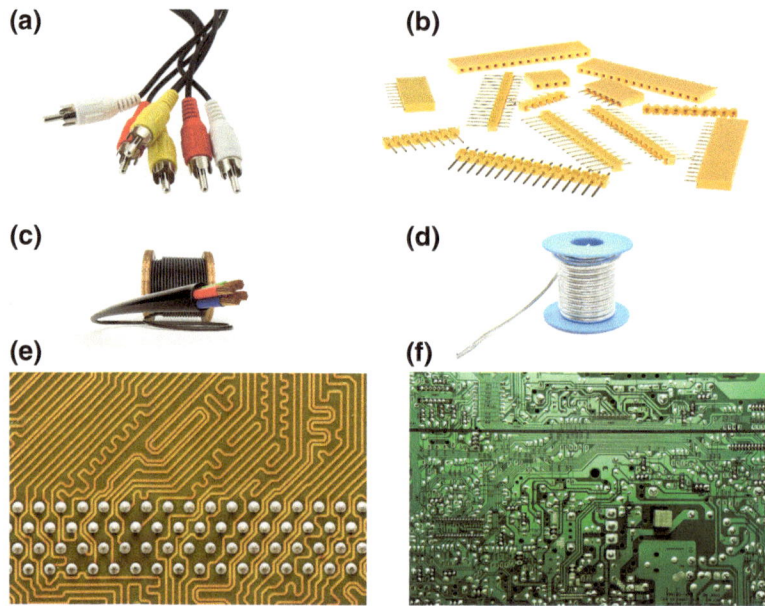

**Fig. 1.8** Examples of forms used to conduct electric current in some technological systems: **a** copper tv cables (iStock.com/thumb); **b** gold connector pins (iStock.com/Teodor Costachioiu); **c** copper wire (iStock.com/Hammad Khan); **d** solder (iStock.com/empire331); **e** copper printed circuit board (iStock.com/Teodor Costachioiu); **f** aluminum printed circuit board (iStock.com/vadimrysev)

tin, copper, silver, and bismuth) is used to connect and conduct current between electrical components. A thin coating of gold is used to conduct electric current in some connectors. Large powerlines frequently use aluminum to conduct electric currents (Fig. 1.8).

## 1.5    Form Variability

### 1.5.1    Modification of Form

The form of an object may be altered or changed to enable the form to provide a particular function. Some form properties are alterable. For example, the shape may be changed. Physical form can be modified, changed, or adjusted. Creating technological objects typically involves change or manipulation of form to achieve a desired function. A paperclip can be straightened and used as a poking tool or bent into a hook. A needle can be magnetized and used as a compass.

The use of stone tools was a fundamental development in early technology. Stones were recognized as suitable for a variety of functions in food preparation, defense, and construction. It was also recognized that the form of some stones could be modified

**Fig. 1.9** Stones and stone arrowheads (iStock.com/thomaslenne)

from initial conditions to facilitate use in cutting applications and weaponry. For example, skillful chipping of stones produced arrowheads such as those shown in Fig. 1.9. Chipping altered the form of the original stones to provide the desired function.

As a modern example, a common type of 3D printing modifies the form of the plastic material by melting and re-depositing through a nozzle. The material retains the same chemical composition, but its shape has been redistributed by the 3D printing process into a form deemed preferable for a particular purpose. The 3D printer can be viewed as a device that primarily modifies the form of the plastic into some other, typically more complex, shape. Figure 1.10 depicts a 3D printer creating a finished part from plastic filament held on a spool.

Modification of form to achieve a desired function is a central focus of modern manufacturing and fabrication. Diverse processes exist for modifying form. These processes include machining from a stock block of material, casting, stamping, rolling, extruding, molding, and forging of metals. Specific examples include: machining a part from stock material, casting a fan blade of titanium, or molding a bottle from raw material. Some of these common manufacturing form-modification processes are seen in Fig. 1.11.

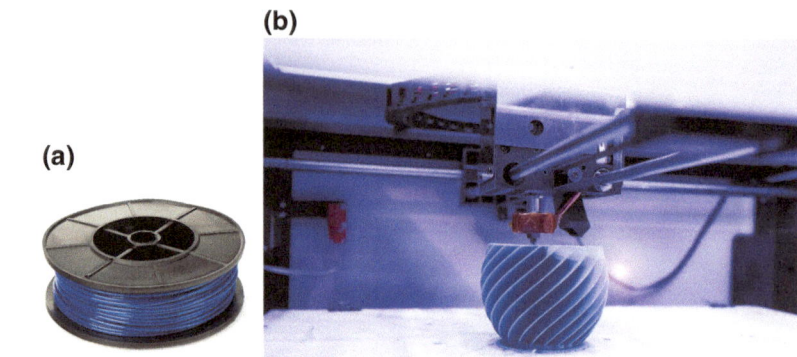

**Fig. 1.10** 3D printer material, printer and finished part: **a** blue polymer filament (iStock.com/Oleh Muslimov); **b** 3d printing machine (iStock.com/kynny)

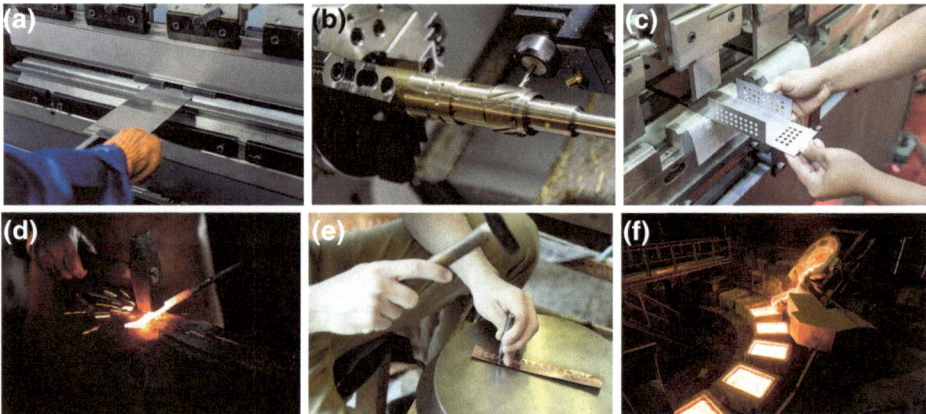

**Fig. 1.11** Examples of processes for modifying form: **a** hydraulic bending (iStock.com/romaset); **b** etching (iStock.com/Phuchit); **c** folding (iStock.com/ake1150sb); **d** forging (iStock.com/grafvi sion); **e** stamping (iStock.com/Coprid); **f** molding (iStock.com/maki_shmaki)

## 1.5.2 Combining Forms

In addition to being modified, forms are frequently combined with other forms to enable the combination to serve a particular function. For example, a common hinge is three separate elements combined to provide the physical support and mobility of a hinged door. Figure 1.12 shows the three elements of a simple hinge.

A transistor is a device that can be used for switching the flow of electric current on and off. Transistors can be made extremely small and a smartphone may contain several billion transistors. Figure 1.13 shows a magnified image of rows of transistors. More than

**(a)**                                                      **(b)**

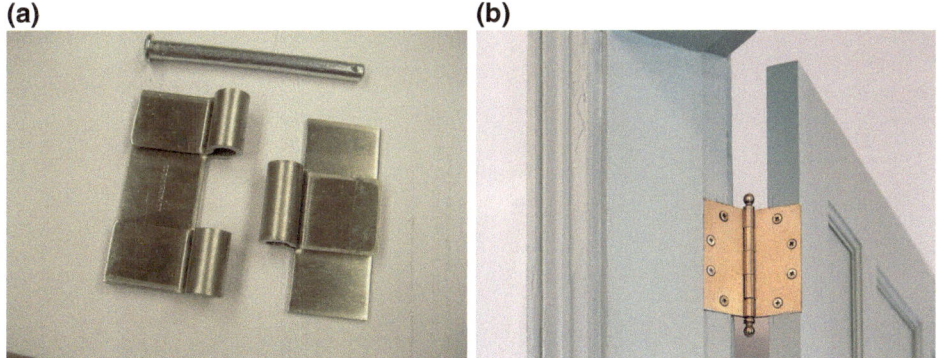

**Fig. 1.12** Common door hinge: **a** hinge in three pieces (iStock.com/Guru3Ds); **b** assembled hinge (iStock.com/aozora1)

**Fig. 1.13** Silicon transistor magnified image [cropped] (iSt ock.com/FroggyFrogg)

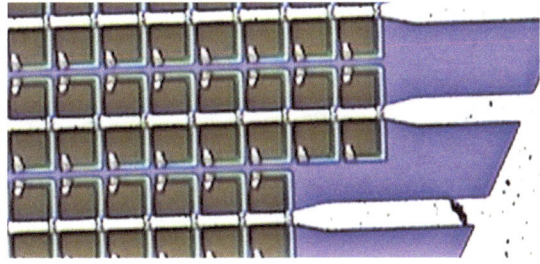

500 of these transistors can fit across the width of human hair. The transistors shown have three main pieces that are made from the element silicon in combination with phosphorous and boron. The silicon is combined with these other elements to achieve desired electrical properties. This is an example of modifying the basic form properties of silicon. Like the hinge, multiple pieces are combined to achieve the final form of the transistor.

## 1.6    Form Properties Enable Function

### 1.6.1    Form Follows Function

Aspects of form determine the ability of an object to carry out or provide a function. A connection exists between form and function. This is sometimes stated as: form follows function. "Form follows function" conveys the idea that a relationship exists between the function of an object and its form. Form properties make functions possible.

Consider a simple example of a rock used to hold open a door. The rock fulfills this function due to its form properties, in this case, mass and the related property of weight

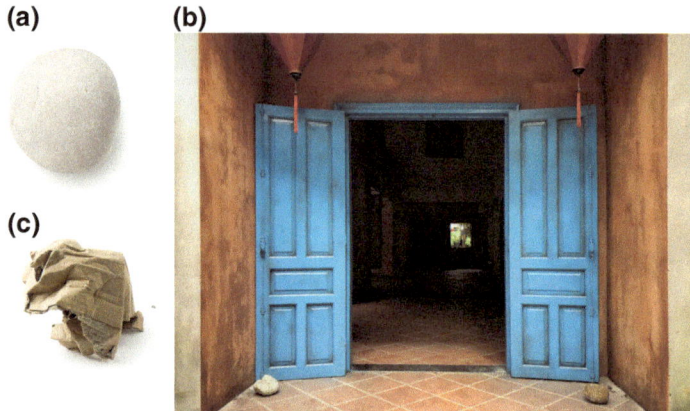

**Fig. 1.14** Rock and empty box as door stops: **a** stone (iStock.com/malerapaso); **b** stone door stops (iStock.com/Coompia77); **c** cardboard box (iStock.com/Altayb)

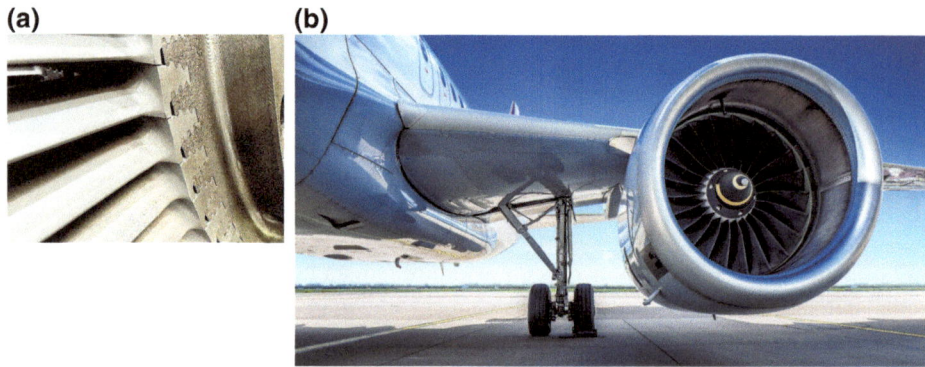

**Fig. 1.15**  Turbine blade and jet engine: **a** turbine blade (iStock.com/Parame Nilrung); **b** jet engine (iStock.com/frankpeters)

combined with the rough surface of the rock. The rock holds the door open. The force closing the door is unable to overcome the friction between the rock and ground surface. The weight and surface texture of the rock enables its function as a door stop. Form enables function. A ball of crumpled cardboard of similar size and shape lacks the weight necessary to function as successfully as a door stop (Fig. 1.14).

As a more complex example of form following function, consider the jet engine turbine blade as seen in Fig. 1.15. When the engine is running, the rotating turbine blades transfer momentum from expanding gasses to rotate the main engine shaft. The form of the blade is highly optimized for its function of interacting with hot gases inside the engine. The blade is made of a combination of nickel, chromium, cobalt, and other metals and has both high strength and high melting temperature to withstand the heat inside the engine. The bottom zig-zag-shaped lower portion of the blade has a profile that allows it to interlock into the main shaft of the engine.

## 1.6.2 The Engineering of Technology Connects Form with Function

Aspects of form determine the ability of an object to carry out or provide a function, that is, to solve a particular problem. This is a key idea of engineering. The form of an object, its physical properties, shape, and other features are determined by what functions the object is expected to provide.

It is the recognition of the interaction between form and function, along with the alteration of form that is central to the creation of technology. The ability to envision a connection between a particular form and the solution to a problem is a characteristic of human ingenuity. The earliest beginnings of technology are characterized by the alteration of naturally occurring objects to solve immediate problems of survival. Much of the body of knowledge of modern engineering disciplines concerns how to specify the characteristics of particular forms to achieve specific desired functions.

## 1.7 Describing Functions in Verb-Noun Form

It would be helpful to have a means for describing function. Functions are thoughts, ideas, or abstract concepts. How are ideas expressed? Ideas are commonly expressed through language using words.

One approach to describing functions uses verb-noun pairs. A verb is a word expressing action or being. Nouns refer to a person, place, or thing. A function is described using an action verb and a noun. Some examples of functions are: support—load, amplify—signal, conduct-current, catalyze—reaction. This verb-noun form can be thought of as illustrating that functions are "some action on some thing." Figure 1.16 summarizes the verb-noun method of describing functions.

The method for describing functions illustrates the transformative nature at the heart of technology. The function of a particular form is described using verbs expressing action. Action implies that some type of change is occurring. The idea of a function is to cause some change in the status of the noun, the person, place, or thing. The central nature of technology, as embedded in the basic concept of function, is to create change in physical reality.

As an example of a process to arrive at a succinct verb-noun expression of function, consider the cup shown in Fig. 1.17. If asked to describe the function of the cup, initial responses might be: "it's used for drinking," "you drink from it," or "it's filled with water." These responses convey aspects of the use of a cup but are not precise descriptions of the

## VERB - NOUN

**Verb**: "word expressing action or being".
**Noun**: "person, place, or thing"

The essence of engineering is effecting change, i.e.
some kind of <u>action on</u> some kind of <u>thing</u>

| (a)            | (b)          | (c)                  |
|----------------|--------------|----------------------|
| Support Weight | Heat Liquid  | Conduct Electricity  |

**Fig. 1.16** Illustration of verb-noun expression of function: **a** weights (iStock.com/abluecup); **b** kettle (iStock.com/PaulPaladin); **c** electrical wires (iStock.com/DustyPixel)

function of a cup stated concisely in verb-noun form. In the verb-noun form, the function of this cup can be stated as: "hold-water."

**Fig. 1.17** Plastic cup holding water (iStock.com/Coprid)

It can be helpful to identify or describe the function of an object in the most general form possible. This practice can help to develop a way of thinking that habitually envisions multiple possible options for solving problems. For example, the function of the cup shown in the figure could be stated as "hold-water." However, a more general way of describing this function could be "contain-liquid." Containing liquid includes the possibility that the liquid contained is water. It should be kept in mind that containing-liquid is just one possible function for which the form of a plastic cup can be used.

Functions can be expressed synonymously. While functions can be expressed in verb-noun pairs, in general, there is more than one way to describe a particular function. Consider the function of a section of pipe. The function of this pipe might be described as channel—flow, guide—flow, direct-flow, conduct-flow. All of these verb-nouns pairs might be used to describe the same action. Language is a diverse tool for expressing ideas. While there are many equally correct ways to state a particular function there are many imprecise or ambiguous statements as well.

Noting that the name of an object is not necessarily a precise indication of function also recognizes the need for a deliberate means for describing function. Since any given form can have a nearly infinite number of potential functions, a challenge in expressing the function of an object in verb-noun form is to describe the function intended in a sufficiently precise manner to avoid misinterpretation. The function of a particular form can be ambiguous or vary from one application to another.

For example, Fig. 1.18 shows some wires. The name "wire" is not sufficient to specify function. Wires are used for many functions. Wires conduct electricity. Wires support loads. Functional descriptions are more specific indicators of use in a particular instance.

Objects can also carry out more than one function at the same time or provide different functions at different times. A light socket both supports and conducts electric current to the light bulb. These functions occur simultaneously. The heater in a drip coffeemaker initially heats the water and later keeps the brewed coffee warm.

Function is an abstract concept or an idea. However once implemented, the functions accomplished by technology are real and affect actual physical objects.

The verb-noun description of function is also seen as a succinct expression of the problem solved or needs to be fulfilled. For example, "support-load" is a statement that a problem exists, some load needs to be supported. Something needs to be held in place or kept from falling. Problems can have many aspects or features, but the verb-noun statement of function summarizes a central nature of the problem. The verb-noun statement of function describes the transformation of physical reality to resolve the problem.

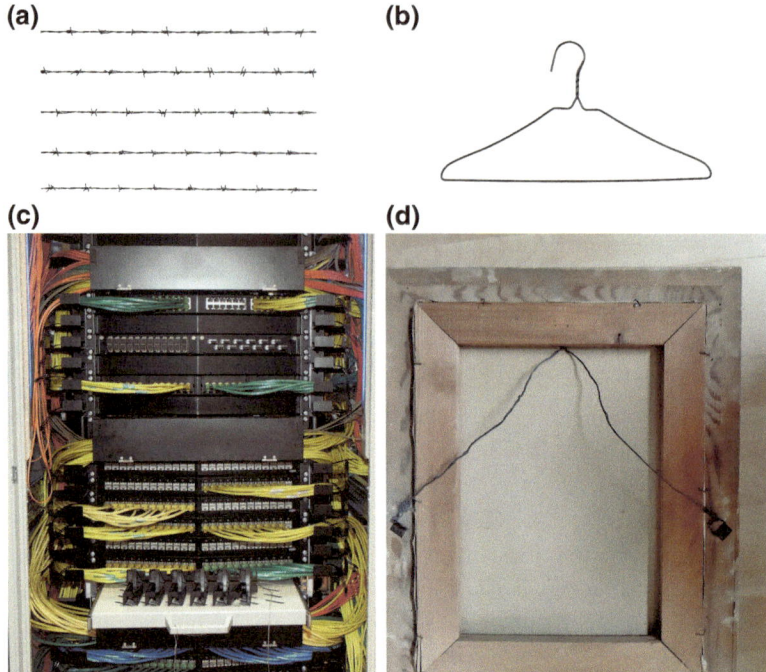

**Fig. 1.18**  Multiple functions of wire: **a** barbed wire (iStock.com/Maravic); **b** coat hanger (iStock. com/DonNichols); **c** wired switch case (iStock.com/jamsi); **d** picture wire (iStock.com/Clarini)

## 1.8    Form Representations and Describing Form

Aspects of function can be expressed succinctly using verb-noun combinations. Describing form can be more challenging due to the diverse range of properties that contribute to form.

### 1.8.1    Physical Dimensions

Technology involves physical objects: a bridge, a car, an artificial knee. There is a need to describe the geometric aspects of form such as size and shape. This is particularly true for objects that do not yet exist. Size, shape, dimensions, and configuration information are needed for the fabrication and manufacture of new technological systems. The form of objects must be described somehow so the object can be made.

It is possible that in some cases the person envisioning a new technological device would not need to describe the form if that person is also the one constructing the device.

However, that designer/fabricator still needs at least some type of mental image of the intended object.

Drawings are the traditional method used to describe the geometric aspects of form and other related information. Many variations exist on the basic idea of a drawing to convey shape and dimensional information. These include Computer-Aided-Design or CAD drawings that provide size and shape information in a planar format. Solid modeling renders an image of an object in a manner that provides the appearance of a 3D object, the view of which can be manipulated. Solid models can be used to create the objects intended using a range of devices including 3D printers of various types and CNC (computer-numerical-control) machining. These formats for representing form all have strengths and weaknesses and are each well-suited to particular applications in the creation of actual physical objects that comprise technological systems. Each is a specialized field of study in the context of the engineering disciplines.

Hand drawing is a traditional method often used by engineers and others to quickly establish a description of form. Hand drawing is immediate, uses simple equipment, and is customary and instinctive. Hand drawing can define the major features of the geometric form of an object. Typically, these drawings are made to scale, retaining the relative proportions and sizes of the dimensions of the object.

Traditional drawing or sketching can be used to convey those aspects of form deemed most significant in a particular application. Drawing presents the opportunity to selectively depict the shape, configuration, and other visual elements that are most important in both identifying the object and in accomplishing the intended function. Photographs while presenting an accurate visual representation can also depict too much detail resulting in an unclear or confusing image. Traditional drawing remains a useful method of representing the form of an object used in a technological system. Figure 1.19 shows the same object rendering using a computer-generated image, a photograph, and a line drawing.

## 1.8.2 Form Material Properties

The form of an object includes the physical and chemical properties of the material from which the object is made. Chemical and physical properties are independent of the shape or volume of the object.

A variety of material properties exist that are useful for characterizing the behavior of materials. Some more commonly encountered properties are listed in Table 1.1. Not all properties necessarily apply to every material. Form properties are related to physical effects. Physical properties comprise a set of form properties. Physical effects include density, melting point, thermal conductivity, electrical conductivity, and yield strength. These effects may be important for particular functions. These types of properties are derived from physical science concepts. Physical properties such as these are described

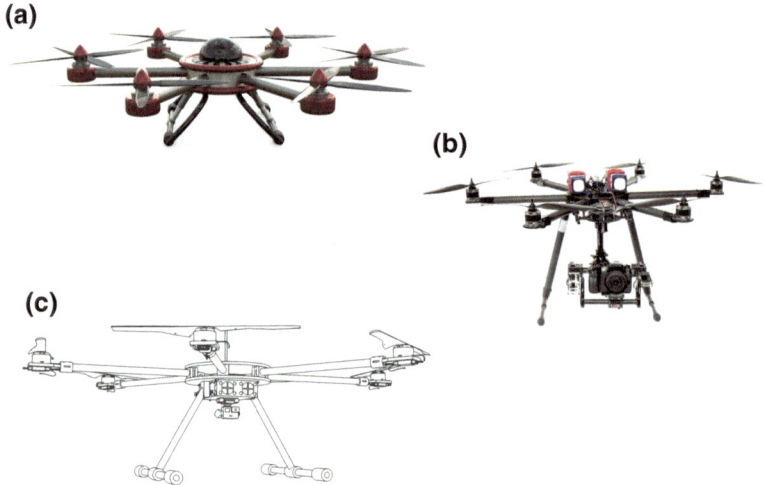

**Fig. 1.19** Methods of representing physical objects: **a** computer-generated image (iStock.com/Vla dimir-H); **b** photograph (iStock.com/MediaProduction); **c** drawing [cropped] (iStock.com/blackl ight_trace)

**Table 1.1** Material properties commonly important in technological applications

| Material properties | |
|---|---|
| Cost per unit mass | Melting temperature |
| Cost per unit volume | Modulus of elasticity |
| Density | Tensile strength |
| Electrical conductivity | Thermal conductivity |
| Hardness | Thermal expansion |
| Heat capacity | Yield strength |

using fundamental physical dimensions and a unit system. Other frequently significant material characteristics are cost per unit volume and cost per unit mass.

The suitability of a material for a particular function often depends on a small subset of the physical properties of a material. Often only a few properties are relevant in a specific application.

### 1.8.3  Form Property Time Dependence

The form of an object can vary with time. Physical dimensions can change for a variety of reasons. Properties can change due to chemical reactions that occur over time. Changes in the environment such as temperature, humidity, or light intensity can change properties.

The effects of use can cause properties to change. The potential of the physical properties of form to change over time may need to be considered in technological applications.

Examples of form changes over time include the rusting of steel exposed to the environment. Tires on automobiles wear down with use. Some rechargeable batteries have a limited number of recharging cycles before they are no longer able to be recharged. Bridges must accommodate the expansion and contraction of the road surface that takes place due to increases and decreases in temperature. Finishes such as paint degrade over time due to chemical reactions facilitated by exposure to sunlight. These are a few examples in which some characteristic of the form changes over time.

Changes in form caused by various means can be unintentional or unwanted such as the rusting of steel or can be deliberate and purposely induced. For example, the heat treatment of metals uses the effect of high temperatures to promote desired property changes in the metal. Even something as simple as allowing wood or glue to dry is an example of a property change deliberately brought about to achieve a more desirable state of the material or object.

## 1.9    Technological Object Recognition

Form characteristics are how people recognize objects. This is clearly stating the obvious, how else would objects be identified other than their physical characteristics, and the definition of form? However, it is a point worth calling attention to because when technological components, not immediately familiar in everyday life are discussed, some means are needed to describe how a particular object might be identified.

Some aspects of the form of an object are readily perceived physical characteristics. Readily perceived aspects of form include shape, proportions or ratio of dimensions, surface texture, specialized features or elements, materials, color, weight, volume, weight relative to size, or configuration of parts. Commonly recognized aspects of form are listed in Table 1.2.

As an example, the object shown in Fig. 1.20 is recognized as a hammer, or more specifically a common claw hammer suitable for driving and removing nails. This object is identified by its form features of size and shape. The hammer is recognized by the characteristic shape of the head, particularly the curved claw. The head made of steel has the common shiny greyish-silver color. The hammer handle has a typical length of about 300 mm (about 12 inches). Hammer handles frequently have a contoured rubber grip extending for about half of the handle length. The handle attaches to the head in the middle between the face (front of the hammer) and the claw. Handling the hammer, the typical mass of approximately 0.5 kg (about 1 pound mass) would be apparent.

Figure 1.21 depicts several objects that resemble the common claw hammer, but closer inspection reveals form features inconsistent with a hammer suitable for driving nails. One is a sculpture with some features similar to a claw hammer but non-functional for

**Table 1.2**  Some properties describing the form of an object

| Readily perceived characteristics |
| --- |
| Shape |
| Size |
| Proportions or ratio of dimensions |
| Surface texture |
| Specialized features or elements |
| Color |
| Weight |
| Volume |
| Weight relative to the size |
| Configuration of parts |

**Fig. 1.20**  Common claw hammer (iStock.com/kyo shino)

driving nails. Two other objects are toy hammers made from plastic or wood. One object is a hammer-shaped cookie. These faux hammers resemble the standard hammer in shape and proportion but do not possess the strength and weight needed for nail driving.

Some of the form features of an object may change over time. For example, an ice cube can melt into a puddle of water. Typically changes in form are due to some type of interaction between the object and some aspect of its surroundings.

Those involved in the design, manufacture, and maintenance of technological systems must be well-acquainted with the form features that facilitate the identification of component elements. In this regard, technology is like every other specialized area of human expertise. Practitioners must recognize specific components or parts of technological systems to a greater level of detail and precision than a typical non-expert. Interpretation of the significance of form features is one aspect of the thinking of experts in any knowledge domain that involves real physical objects. A pathologist should be able to recognize

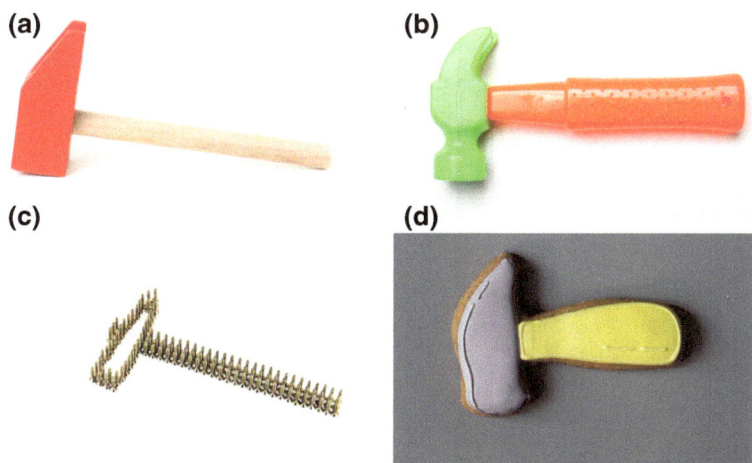

**Fig. 1.21** Faux hammers: **a** wooden hammer (iStock.com/real444); **b** plastic hammer (iStock.com/realrrr); **c** hammer artwork made with galvanized screws (iStock.com/GeorgeK); **d** cookie hammer [cropped] (iStock.com/Natalya Mikhalkina)

cancer cells in a biopsy. An engineer inspecting a domestic solar electric system needs to recognize the DC-to-AC inverter that converts direct current DC from the solar panels to alternating current AC required by home appliances (Fig. 1.22).

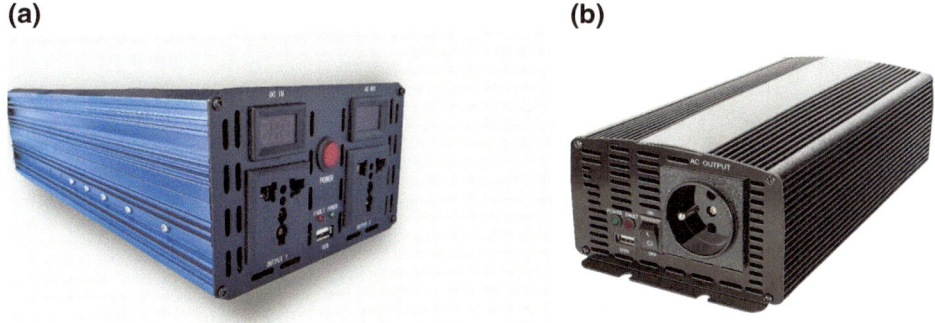

**Fig. 1.22** DC to AC Inverters: **a** (iStock.com/Weeraa); **b** (iStock.com/kuczin)

## 1.10    Visual Representation of Function

An expression of a function is an expression of an idea. Visual representations of functions describe ideas rather than an actual physically accurate reproduction of an object's appearance. Representations of functions use symbols. This is not surprising because symbols are often used in human communication to represent more complex ideas in an abbreviated form.

Representation of function can take the form of schematic diagrams. These diagrams use symbols that represent a particular established component that provides a known function. The symbol provides a quicker means of depicting the component than a drawing intending to be visually accurate. Specifically established symbols for particular components allows practitioners to interpret each other's diagrams. Just like any language, communication is facilitated to the extent that the elements have a shared interpretation.

Schematic diagrams for electrical systems are one of the most well-known functional representations. An example is shown in Fig. 1.23. Standard, easy-to-draw symbols exist for commonly used components. The components provide well-defined and established functions used in the device depicted in the given schematic. Symbols representing function are conveying an idea rather than a specific physical shape or dimensions.

The next figure, Fig. 1.24 shows a complex circuit. The circuit has numerous components and interconnections but the basic approach of using symbols to represent functions is the same as in a simpler circuit. This diagram is not a physical representation of the

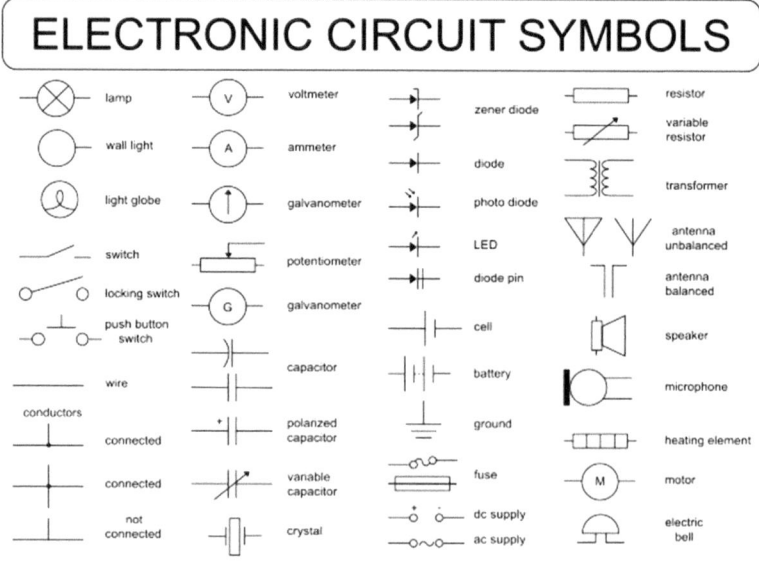

**Fig. 1.23**  Electrical component symbols (iStock.com/frentusha)

**Fig. 1.24** Complex electrical schematic and corresponding physical depiction (iStock.com/Mercedes Cavani)

particular device. The schematic diagram shows the transformations taking place in each component and the steps that occur in the overall process at the level of the individual components.

While the symbols are representing a particular function that can be accomplished in many possible ways, some function symbols might evoke the physical aspects of the form used to accomplish the function. In piping systems, a common function is the need to control the rate of flow. A valve is a common component used to control flowrate. Figure 1.25 shows a typical valve and a symbol for a valve. The valve symbol bears a highly simplified resemblance to the physical form of a valve.

Functional visual representations show interactions. There is at least one input and one output. Symbolic visual representations of functions tend to include some indication of a connection between that element and other elements external to that component. The electrical and pipe system schematic symbols all indicate that the functional elements connect to something else. Form representations are focused on the physical nature of the object itself. Function representations emphasize the role or purpose of the object as it interacts with other components in the system.

The most general representation of a function is simply a box with a description of the function. This is sometimes referred to as a block diagram representation. Figure 1.26 illustrates a basic block diagram representation of function. This block diagram depiction

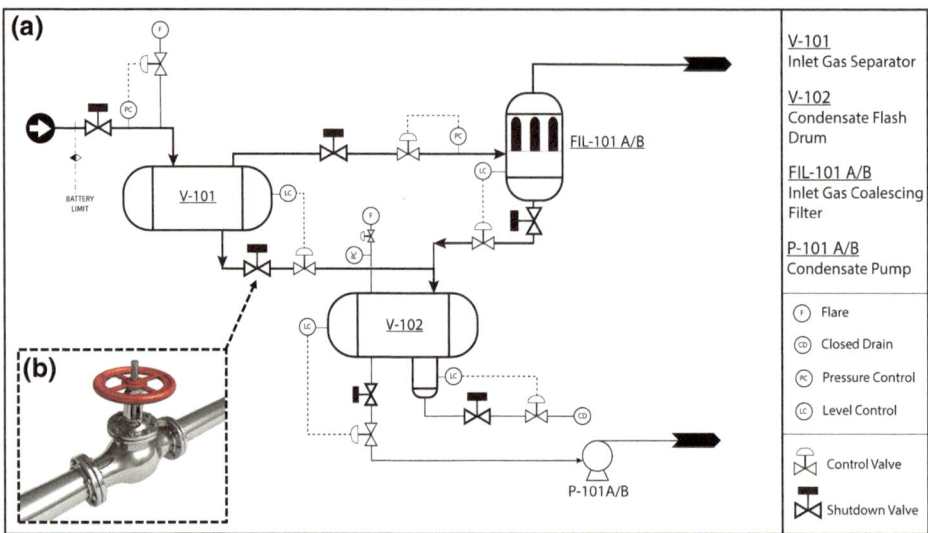

**Fig. 1.25** Valve and valve symbol: **a** diagram with valve symbols (iStock.com/Mercedes Cavani); **b** valve (iStock.com/scanrail)

**Fig. 1.26** Block diagram representation of function

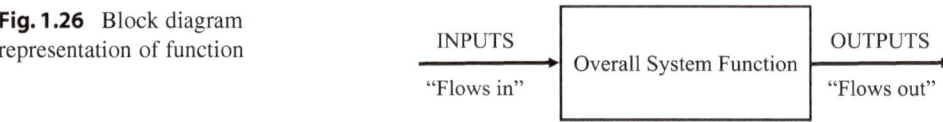

of function puts the focus on the transformation occurring at that point and not any details of the object used to accomplish this functional transformation.

The use of the block diagram representation can help acknowledge that there is more than one way to achieve a function. The box includes no form information about the means used to achieve the function. This emphasizes that the desired function is not necessarily restricted to implementation via any one particular form and the possibility exists that it might be achieved in any number of possible ways.

## 1.11   The Problem–Function Relationship

The existence of a problem implies that some kind of change or transformation is needed. The technological system's function is the transformation enabling problem resolution. The function is the actual problem solver.

What is the relation between technology as problem-solving and the functions that can be carried out by a physical object? Function is the problem solved by a particular form.

**(a)**            **(b)**                        **(c)**

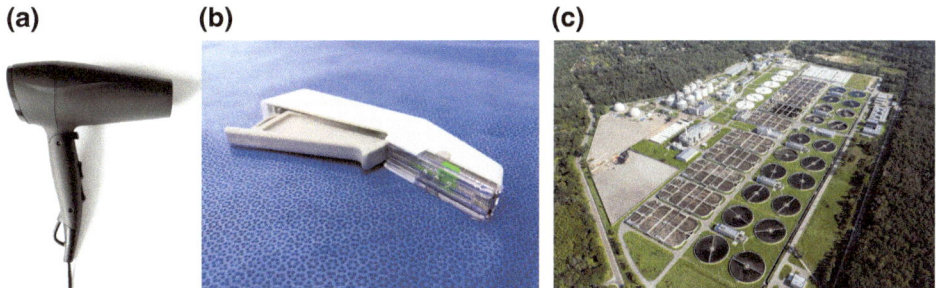

**Fig. 1.27** Problem solutions: **a** hair dryer (iStock.com/olhakozachenko); **b** skin stapler (Sen Alfred/Shutterstock.com); **c** wastewater treatment plant (iStock.com/Bim)

What is a problem? A problem is a situation needing change. If change was not needed there would not be a problem. The solution to a problem requires some type of change.

The functions carried out by technological systems are solutions to problems. Functions solve problems as the action leading to change. Change is central to the concept of function. Verb-noun pairs describe functions. A verb is an expression of action. A noun is a person, place, or thing. A function is seen as some action on some thing. To solve a problem, a change of condition is needed. In other words, some kind of action on some kind of thing is needed to solve a problem.

Three examples are shown in Fig. 1.27. All of these items are part of our technological landscape and can be viewed from one perspective as solutions to problems. The familiar hairdryer solves the problem of wet hair, its function is to dry hair or more precisely remove water from hair. The skin stapler is a biomedical device whose function is to staple skin. It solves the problem of closing wounds. The function of the wastewater treatment plant is to remove pollutants from water. It solves the problem of purifying wastewater. In each example, it is the action that resolves the problem: dry, staple, purify.

## 1.12  Value

What is value? For technological systems, their value is related to the importance of the problem solved. Not all problems are of the same significance. Some problems are more important than others. The value of a solution is proportional to the importance of the problem. How bad is the problem? Value reflects the desirability of the solution. The value of the solution is an indicator of the urgency or priority of the problem.

The value or the importance of the problem is reflected in the degree to which those paying for the solution are willing to allocate their money, time, or other resources to obtain that solution. Generally, the more critical the problem or the need, the more resources the consumer is willing to devote to obtaining the solution. Typically, the cost or price is the main determiner of value, however, consumers have other assets that they might choose to use to help obtain something that they want. For example, time or personal influence might be considered assets a consumer might employ to obtain something they desire. It is the customers purchasing technological systems who define the value of the solution.

Value or importance can be perceived value. The customer allocates resources based on the perception of importance. Importance is a subjective judgment on the part of the consumer. Perception can be influenced by numerous factors including advertising, social groups, and cultural practices. Importance may also depend on circumstances and can vary over time.

## 1.13  Producer–Consumer Disconnect

In the current era, consumers of technological systems usually do not make technological items for themselves. Modern technology production involves complicated systems which produce and deliver technological solutions to the user whether it is a paperclip or a jet. Technology consumers of all types purchase the technology from various entities that produce it. For example, people purchase their smartphones from producers like Apple and Samsung. Because consumers generally do not produce their own technological systems, but rely on specialists who create technology for other people, there is often a disconnect between producers and consumers of technology.

## 1.14  Who is the Customer?

Who is the customer? Who is the problem owner? Generally, the customer is the end-user of the product or the consumer of the technology. The idea of an end-user can be complex in a non-trivial fashion. Many technological products can be seen as having multiple groups of consumers. The needs of all groups of consumers must be met simultaneously.

For example, consider a medical device such as an arterial stent. Stents are often used to open narrowed coronary arteries. Who are the consumers of this technology? Who has the problem to be solved? Ultimately the stent is installed in a patient with a narrowed artery so the end-user could be the patient that receives the stent. Another consumer or user that has a problem solved is the doctor or person providing treatment. Successful installation of the stent and effective treatment of the patient is also the doctor's problem. Looking more broadly the costs of this patient's procedure may be covered by some type of health insurance. The insurance organization is likely paying for the procedure. The insurance company might be considered a customer as well since they are involved in providing payment and have an interest in the patient's outcome. It is possible to identify multiple different types of consumers of the arterial stent that value the function of opening the artery.

## 1.15  Technological System Design Requirements

Design requirements are the link between the consumers and producers of technology. Because the end-users are not the producers, the development of technology must be guided by a set of requirements that describe the problem to be solved and define the characteristics of an acceptable solution.

### 1.15.1  Defining the Problem

An initial phase in the development of a technological system to solve a problem or meet a need is to specify what the system must do. Problem definition is the first phase of the process of designing and creating technological systems. The problem is defined in terms of solution requirements. The technological system solves the problem if it can meet all of the requirements.

The problem definition process requires input from the customers and end-users. To solve a problem for other people, technology developers must understand the nature of the problem from the perspective of the customers and users. The customers ultimately decide whether or not a particular technology solves the intended problem.

In the case of widely used consumer products, customer needs may not be uniform. In addition, nearly every technology has different categories or types of users that may perceive the problem differently. Different sub-groups of users may have different requirements. The technological solution may need to accommodate a diverse range of solution requirements.

### 1.15.2  Customer Defines the Solution

Who determines what is an acceptable solution to a problem? The development process creates technological systems that solve individual and societal problems. It is the various customers that ultimately decide whether or not a particular technology solves the intended problem. The customers assign value to the problem and determine if a solution is acceptable.

This situation could be likened to a physician treating a patient. Ultimately the final condition of the patient determines the success of the treatment. Does the patient recover? How does the patent feel? If the patient's health improves the medical treatment is successful. Similarly, the customers decide if a technological solution is successful in addressing a particular problem or need.

The customers decide if a particular technology is a success in solving a problem, but the technology is not created by the consumers, it is created through a development process carried out by other people. The development of technology must be guided by a set of requirements that describe the problem to be solved and define the characteristics of an acceptable solution.

The process of determining the requirements for a mass-market technological product is a complex one toward which technology companies can devote considerable resources and effort. The intent here is to describe an overview of major aspects of the nature of technological system requirements. As with most areas of technology different subfields, industries, and companies, will have different approaches for understanding the problem that is optimized for that circumstance.

### 1.15.3 Do and Be

Design requirements might be considered of two types: what the product must do and what the product must be. Technology solves problems so the product must "do" something to solve the problem. What the product must "be" describes attributes of the product that the consumers require as part of an acceptable solution.

### 1.15.4 Do-Functional Requirements

An initial step is to determine what the system must do. These are functional requirements. Functions are the transformations that solve the problem. Functions are high-level requirements that convey the major behaviors and actions accomplished by the technological system.

The major functions of the system are the problem solutions. Functions are abstract. Functions are ideas. A function is not a thing. Functions are transformations of inputs into some outputs expressed in verb-noun form.

There is a tendency for people to express system requirements as a thing needed rather than a function. This is natural since we are more accustomed to thinking in concrete terms, in terms of objects rather than abstract ideas. In problem definition, an effort must be made to extract functional needs from needs embedded in descriptions of things. Going beyond the thing to the need satisfied is important. Functional requirements describe what the system must do not what the system must be. Complex technological systems by nature can involve a complex series of functions.

### 1.15.5  Be-Form Requirements

Some aspects of the solution are expressed in terms of the physical characteristics of the technological system. These types of requirements define the form aspects of the system and put constraints on the forms used in the solution. An example of a form requirement is "it must be blue" or it must fit in a box that is 100 mm on each side or a particular motor must be used. In some cases, existing standards, codes, or laws will lead to particular features in the form of a solution. For example, if a technological system will be plugged into an electrical outlet in the United States, then a particular type of plug with pre-determined dimensions should be used.

In general, particular form requirements limit the possible solutions to the problem. Recall that a given function can be accomplished in many possible ways. Excessive or inappropriate form requirements can limit potential approaches to accomplishing the desired functions.

Functional requirements describe what the system must do not what the system must be. Functional requirements do not inherently limit the designers' options. Form requirements in some sense limit the designers' choices because they effectively determine some aspect of the system design and restrict flexibility.

## 1.16  Engineering Specifications—Measurable Requirements

In developing technological systems there is a need to be able to establish that an acceptable solution has been achieved. There must be a way to unambiguously demonstrate that the requirements have been satisfied. This is particularly important considering that the end-users of technology are not the producers. The producers need to be able to show that they have met the design requirements.

Specifications are design requirements stated in measurable ways. If the requirements are measurable then an independent entity can examine a technological system and determine or verify that the design requirements have been achieved.

Requirements on the form of the technological system are by definition measurable. Form is a physical characteristic. For example, if a system is expected to be lightweight, what defines lightweight? Weight is a form property, so it can be measured in some way and a quantified target goal can be determined. For example, weight should be under 20 pounds (or equivalently, 9 kg).

Functional requirements must have quantified performance specifications associated with them. While a function is an abstract idea expressed in verb-noun form, the ultimate implementation must result in a transformation in the physical world. For example, if a technological system is needed to convert mechanical power into electrical power, then some quantified amount and rate of energy conversion are needed expressed in appropriate physical dimensions. In this case Watts. The input and output are measurable physical quantities. So an example might be to convert 20 kW of available mechanical power to at least 5 kW of electrical power. Note that the specification does not say anything about how the conversion is accomplished. The issue of concern is the amount of output power and the available input power. Functions can be accomplished in multiple different possible forms. In the case of specifications on functions, any "how" that achieves the desired measurable transformation of inputs to outputs (and meets any other relevant requirements) is acceptable.

While functional requirements may be stated in abstract terms and accomplished by many potential means, the quantitative specifications on functions establish actual physical indicators that the technology is achieving desired goals. Specifications on function describe the changes in the physical world achieved through the use of the particular technology.

Technology changes the actual physical conditions in which people live. It is necessary to express the solution requirements in quantified terms to establish that a solution has been achieved. The needs of the customers must be expressed in some form that can be measured or otherwise quantified to be able to establish objectively that the final product solves the problem. Expressing solution requirements in measurable, quantitative form enables objective verification that the needs of the consumer have been met.

## 1.17  Overview: Form, Function, Value

This chapter started with the idea that physical objects can be viewed from the perspective of form and function. A given form can be used for a variety of functions and any particular function can be accomplished by multiple forms. Technology solves problems creating a change in the physical environment. The problem solution is a transformation from an existing condition, available inputs, to a transformed physical reality with the desired outputs.

The importance of the problem is reflected in the extent to which a consumer of the technology is willing to allocate their various resources to obtain the solution. The importance of the problem determines the value of the solution. Problem importance and value are determined by the users of the technology, not the producer.

Because generally the consumers of technology are not the creators, some approach is needed to communicate the needs of the consumer and the nature of the problem to the producers of the technology. This communication is accomplished by developing a set of design requirements. Design requirements address the two perspectives on physical objects: form and function. Functional requirements focus on the changes or transformations that the system must accomplish. Form requirements are conditions, often viewed as constraints, on the physical characteristics of the solution. To validate that the technological system solves the desired problem, requirements must be quantified into measurable specifications. The physical properties of form are inherently measurable.

Functional requirements are made measurable in terms of quantification of the inputs and outputs.

The creation of technological systems are changes to physical reality. Design requirements are specifications that determine the physical features or form of that new reality and the impact of that technology on the existing world in terms of the amount of the existing inputs that are converted into outputs by the functions occurring in the system.

## Bibliography

Akiyama K., *Function Analysis: Systematic Improvement of Quality and Performance*, Andrew P. Dillon, Translator, Productivity Press, Cambridge, (1991).

Cloutier, Robert, Clifton Baldwin, and Mary Alice Bone. *Systems Engineering Simplified*. 1st edition. CRC Press, 2017.

Eggert, Rudolph J. *Engineering Design*. Upper Saddle River, N.J: Pearson College Div, 2004.

Hatley, Derek J., Peter Hruschka, and Imtiaz A. Pirbhai. *Process for System Architecture and Requirements Engineering*. 1st edition. New York: Dorset House, 2000.

Otto, Kevin, and Kristin Wood. *Product Design: Techniques in Reverse Engineering and New Product Development*. 1st edition. Upper Saddle River, NJ: Pearson, 2000.

Ullman, David. *The Mechanical Design Process*. 4th edition. Boston: McGraw-Hill Education, 2009.

Ulrich, Karl T., and Steven D. Eppinger. *Product Design and Development, 4th Edition*. 4th edition. Boston: McGraw-Hill, 2007.

# Systems Transforming Materials, Energy, and Information

**2**

## 2.1 Chapter Overview

- A system is a group of objects, forming a network, to achieve a common purpose.
- A technological device regardless of size or complexity can be considered to be a system.
- A system is identified by an imaginary boundary that encompasses or encloses the system separating it from everything else in its environment. The system interacts with its environment by inputs and outputs that are identified at the system boundary.
- The system function is recognized by the transformation of inputs into outputs.
- Inputs and outputs of a technological system can be classified as material, energy, or information.
- Material refers to matter having the property of mass.
- Energy is the ability to do work. Energy exists in various forms such as kinetic, potential, and chemical. Power and force are related to energy and may also appear as system inputs and outputs.
- Matter and energy are conserved.
- Information, strictly speaking is energy, in the form of data or signals and is given a special designation in a technological system.
- Some engineering fields can be distinguished by the types of transformations of energy, materials, and information with which they are primarily involved.
- Different modes and phases of system operation can involve different inputs and outputs.
- Technological systems are frequently characterized by concepts of efficiency defined by a ratio of desired outputs to necessary inputs provided at some cost to the user.

© The Author(s), under exclusive license to Springer Nature Switzerland AG 2024    33
J. Krupczak, Jr., *Understanding Technological Systems*, Synthesis Lectures
on Engineering, Science, and Technology, https://doi.org/10.1007/978-3-031-45441-7_2

## 2.2    Definition of a System

### 2.2.1    The System Concept

A system is a combination of interacting elements or parts forming a more complex whole that is considered as a single unit. Some examples of systems are a subway system, the circulatory system and a river system (Fig. 2.1).

A river system is a network of streams, lakes and tributaries that merge together into a larger river. What is called the Washington D.C. Metro consists of all of the many tracks, stations, railcars, and other elements that transports an average of 500,000 people each day. The human circulatory system is comprised of the heart and a network of arteries and veins that transport blood throughout the body.

The components of a system form a more complex whole. To achieve a whole more complex than the pieces, the individual system elements must interact in some way. There must be some type of exchange between some components of the system. Systems are characterized by dynamic interactions between the elements. For example, in the Metro system the passengers, railcars and stations interact. The passengers enter the stations and at the stations board various railcars. The railcars enter and leave stations acquiring and discharging passengers.

Without interaction between elements the system is simply a collection of objects. A collection is a group of items that do not form a more complex whole. The exchanges between components are essential to the identification of a system. The opposite of a system is a group of unrelated objects. Figure 2.2 illustrates the idea of a collection as distinct from a system.

The un-relatedness that characterizes a collection is based on the lack of interaction between the components. The collections of objects shown in Fig. 2.2 do not interact

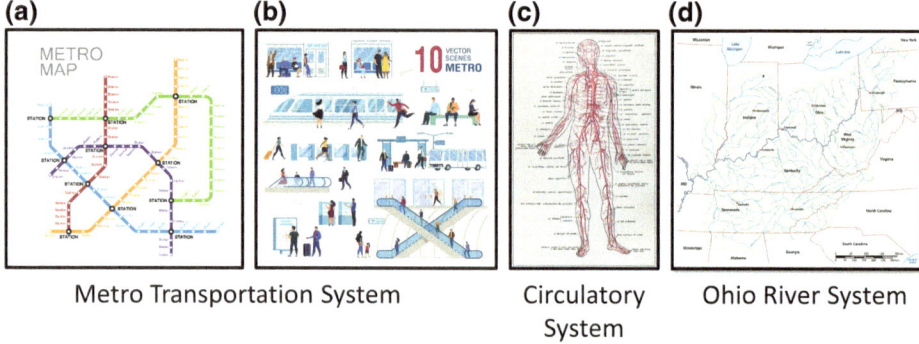

<table>
<tr><td>(a)</td><td>(b)</td><td>(c)</td><td>(d)</td></tr>
</table>

Metro Transportation System          Circulatory          Ohio River System
                                        System

**Fig. 2.1** Examples of systems: **a** metro map (iStock.com/maximmmmum); **b** metro scenes (iStock.com/Tera Vector); **c** circulatory system (iStock.com/ilbusca); **d** Ohio river system (iStock.com/Rainer Lesniewski)

**Fig. 2.2** Collections of elements not forming a system (iStock.com/Johnrob)

in any perceptible way. No more complex whole emerges from the combination of the group.

The Metro system is no longer a system when the dynamic interchanges are absent. Imagine the metro system with no passengers and with the railcars motionless on the tracks. This former Metro is nothing more than a collection of interesting objects: buildings, steel tracks, empty railcars here and there. Without the dynamic interactions between elements the system simply becomes an unrelated group.

### 2.2.2  System Boundary

The concept of a system boundary is used to distinguish a system from everything else in its surroundings. The system boundary is envisioned as an imaginary enclosure that completely contains the system. The boundary separates the system from its environment. The boundary may not be an obvious physical object. In some cases, the boundary must be artificially designated.

For example, the idea of a Metro system is readily grasped; however, the actual boundary or beginning and ending of the Metro is much less obvious. Where does the Metro begin and the city end? For example, is the start of the Metro the sidewalk in front of the station or is it the first step of the entrance stairs?

Because the natural boundary between a system and its surroundings may be unclear it can be necessary to designate what separates a particular system from everything else. In the case of the Metro an imaginary boundary can be drawn around all the elements of the Washington D.C. Metro system. The boundary encloses the tracks, stations, and railcars of the metro. This boundary serves to distinguish what is considered to be the Metro from the rest of the surrounding environment of the city.

The need to specifically designate a boundary of a system is comparable to the process of establishing the boundary between legal entities when no easily recognized natural boundary exists. An example is the boundary between states. Michigan and Indiana illustrate this point. There is no natural physical indication of where one state ends and the other begins. The concepts of the state of Michigan and Indiana are superimposed on an otherwise continuous naturally occurring landscape. Nevertheless, at the present time, these two regions are distinct from a human perspective. The states have different laws, tax rates, and civil infrastructures, so some type of boundary between one system and the other must be established.

### 2.2.3  System Inputs and Outputs

In addition to internal interactions, systems interact in some way with the rest of the world. A system is connected and interacts with its environment by inputs and outputs that are identified at the system boundary. System inputs and outputs are identified at the system boundary.

Passengers enter the Metro at one station and exit at another. Oxygen from the air enters the circulatory system at the lungs and carbon dioxide leaves. Water from precipitation flows into the river system at tributary streams then flows to the river before leaving the system by flowing into the ocean.

Implied in this brief look at system inputs and outputs is the idea that some type of time frame is used when identifying inputs and outputs. Systems are dynamic and the status of the inputs and outputs can vary at different times or for different aspects of system behavior.

The Metro has more passengers entering and exiting at some times of the day. Outlying stations have more passenger input during the morning and more passengers exiting in the evening. The amount of precipitation entering a river system varies with weather conditions leading to increasing and decreasing flow downstream sometime later. The rate of oxygen input and carbon dioxide output of the circulatory system will be different when the body is exercising.

## 2.3  Technological Systems

### 2.3.1  Technological System Definition

Nearly all of the technological products that comprise our human-built environment can be considered to be systems. The term *technological system* can be applied to most technological devices regardless of size or complexity. A technological system is a "combination of interacting elements organized to achieve one or more stated purposes[1]".

---

[1] International Council on Systems Engineering (INCOSE) Systems Engineering Handbook v3.2 (2010).

**(a)**          **(b)**                                    **(c)**

**(d)**

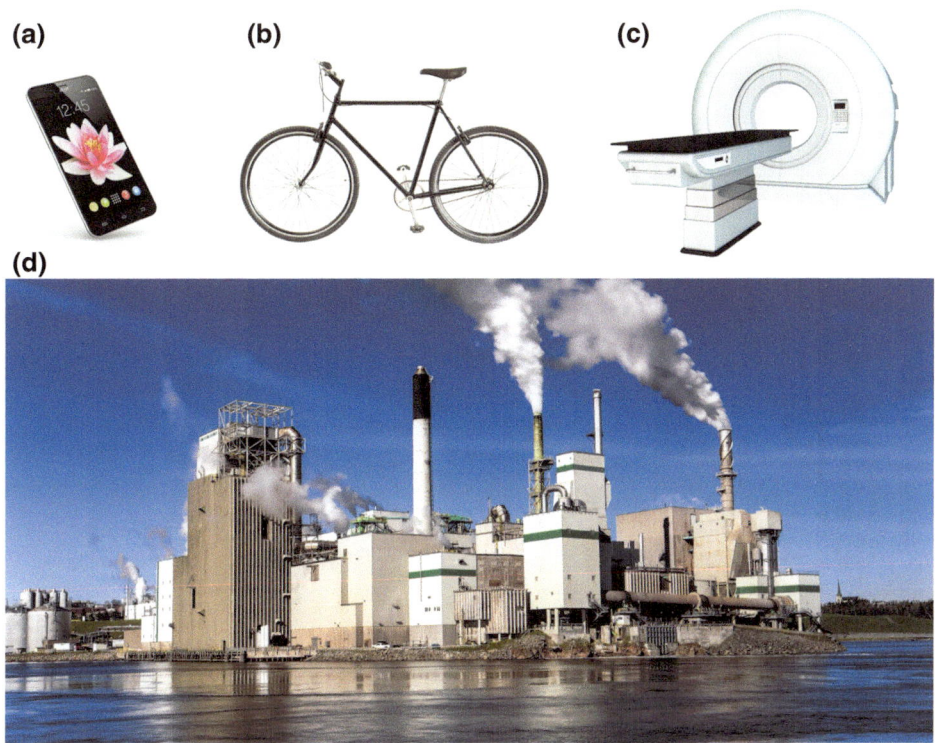

**Fig. 2.3**  Technological systems: **a** mobile phone (iStock.com/scanrail); **b** bicycle (iStock.com/Rin oCdZ); **c** MRI scanner (iStock.com/3alexd); **d** paper mill (iStock.com/SergeYatunin)

The term technological system will be used to refer to any engineering product or object regardless of size or complexity. A component, device, assembly, factory, or structure can be considered as a technological system. The perceived complexity of objects is not the distinguishing feature that determines what is considered a system. Some examples of technological systems are depicted in Fig. 2.3. While clearly a mobile phone with billions of separate transistors must be treated differently than a bicycle in some types of analyses, the system framework is helpful as a unifying structure for understanding and describing both of these devices.

## 2.3.2   Technological System Function

Engineered technological systems are created for some particular purpose or function. These systems are created to solve problems. Solving the problem is the function of the system.

To be a "system" the components of a technological system must interact in some way to achieve a whole that is more complex than a collection of pieces. The dynamic nature of technological systems is the transformation of some type of inputs into some type of output. Something enters the system by crossing the system boundary, something happens within the system due to interactions of the components and some type of outputs leave the system by crossing the system boundary.

Creating some type of change is a central purpose of any technological system. A technological system transforms a set of available inputs into some type of desired output. Technological systems are transformational; they change existing conditions into a desired output state.

The basic system concept of inputs and outputs is illustrated in Fig. 2.4. Consider a smart phone. For the end user some of the problems or needs solved by this device are communicating with others, accessing information, and providing entertainment.

The consumer expects the device to carry out particular functions. In this case the functions can be described as sending and receiving text messages and phone calls, sending and receiving information from the internet, and running applications. The essential output of all of these operations is some form of visual display information along with possibly some type of sound.

To provide these functions certain inputs are required. These are: electric current to charge the internal battery, data in the form of signals containing the text, audio, and video information. Other inputs are the user selections or operations of the device control panel.

**Fig. 2.4** Mobile phone (iSt ock.com/scanrail)

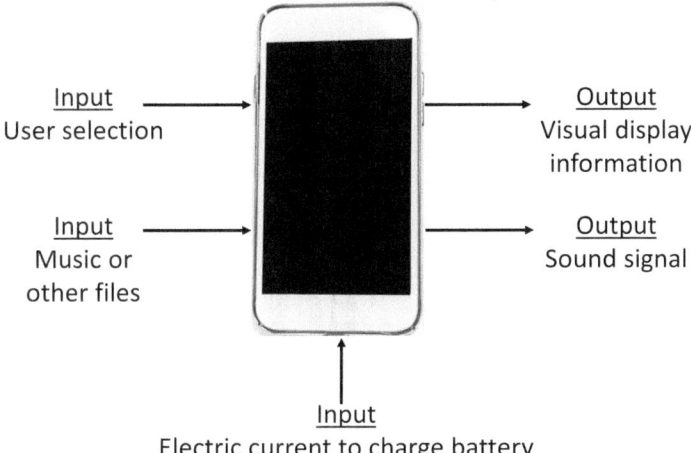

**Fig. 2.5**  End-user view of a technological system (iStock.com/loops7)

The device produces outputs in the form of an output signal that can be converted to sound via headphones or other suitable equipment and output in the form of information appearing on the visual display of the device. Outputs can include signals that communicate with other devices.

This describes the basic consumer-level interaction with the technological system. Particular functions are expected. The user expects to provide certain well-defined inputs and receive expected outputs (Fig. 2.5).

### 2.3.3  The Block Diagram Model

This description of a technological system as providing a function in converting well-defined inputs into outputs is sometimes called the "Block Diagram" model. In the "Block Diagram" approach the details about how inputs are converted into outputs are not specified, and unknown by the end-user. The input disappears into the "box" that somehow produces the output.

Figure 2.6 depicts the essential aspects of the block diagram model. The overall system function expresses the main or most important function of the system. This is what the end-user or consumer expects the product to do—first and foremost. The overall system function is expressed in verb-noun form.

The inputs and the outputs can be considered as flows into or out of the system. The inputs and outputs need not be just physical objects. Inputs and outputs can be such things as sound, heat, electric current, or light.

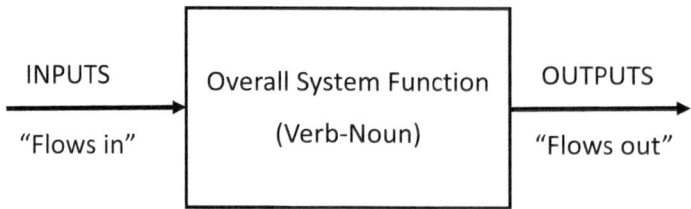

**Fig. 2.6**  The "Block Diagram" representation of a technological system

In the block diagram model, the end-user or the consumer is viewed as expecting to provide well-defined inputs and receive the desired outputs. In the block diagram model, the means by which the inputs are transformed into the outputs are not specified. The system is a box in that it is conceptually opaque. The end user is unaware and unconcerned about what exists inside the box so long as the desired outputs are produced.

### 2.3.4  Technological System Boundary

When analyzing technological systems, as is the case with any type of system, it is necessary to define or establish a boundary between the technological system and its surroundings. Everything not part of the system is part of the system's surroundings or environment. A demarcation must be clear between what is within the technological system and what is part of the external environment or outside the system. The system boundary distinguishes between what is viewed as part of the technological system under consideration and what is not.

The system boundary is an important concept because the boundary defines the system. One definition of a system is: "*an entity connected to its environment by means of inputs and outputs identified at its boundary.*[2]" The system boundary designates what is to be considered part of the system and what is part of the environment. Environment does not necessarily mean just a natural ecosystem but rather anything that is not the system under consideration. The environment of one technological system could include other technological systems. Also, the inputs and outputs to the system can be recognized or specified only at the system boundary.

The system boundary does not necessarily have to correspond to a physical entity. It is an imaginary construct for the purposes of clarifying the analysis. In many cases the system boundary may be coincident with the physical boundaries of a device. For example, in the case of the smartphone device in Fig. 2.4, the outer casing of the device could serve as a natural and logical location for defining the system boundary.

---

[2] Otto and Wood ( 2000).

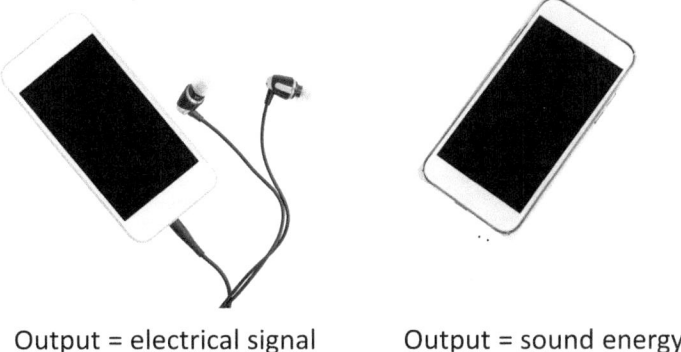

Output = electrical signal        Output = sound energy

**Fig. 2.7**  Source with and without earphones **a** smartphone with earphones (iStock.com/Valeriia Soloveva); **b** smartphone without earphones (iStock.com/loops7)

Consider a different technological system—a utility electric power system. If the electric power distribution grid is the technological system under consideration, then establishing the system boundary is less obvious. For example, is the system boundary at the power plant where the plant output connects to the distribution lines? An alternative boundary for the system might be the point at which an individual end user such as a house connects to the distribution wires. It might even be considered that the internal wiring of the house is part of the distribution system, and the system boundary should extend to the electrical outlet into which individual devices using electricity are connected. The system boundary depends on the intent of the analysis.

The choice of a system boundary is not necessarily trivial and can have an impact on what are the inputs and output of the system. Consider the smartphone shown in Fig. 2.7. If headphones are considered part of the system, then sound is one output of the system. If the system is considered to be just the device itself, then the same output might be a Bluetooth radio signal.

The system boundary can be indicated by a dotted line drawn around the system. This establishes the technological system under consideration. Everything else constitutes the external environment in which this system operates. Figure 2.8 illustrates a dotted line box used to designate the system boundary. In one case the earphones speakers are part of the system. Sound crosses the boundary. In the other case the speakers are outside the system and an electrical signal crosses the system boundary.

### 2.3.5  Technological System Function Stated in Verb-Noun Form

The overall function or functions of a technological system are expressed in verb-noun form just as any other function. The verb-noun combination describes some type of action

**Fig. 2.8** Dotted line used to designate boundary of system: smartphone with earphones (iStock.com/Valeriia Soloveva)

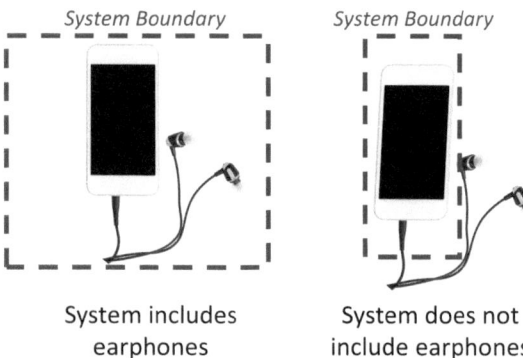

System Boundary          System Boundary

System includes          System does not
earphones                include earphones

(the verb) on some thing (the noun). The verb-noun description of function conveys the transformation or change that is brought about by the system. In the case of the smartphone some of the functions are display video images, or output a sound signal.

## 2.4    Materials, Energy, and Information

The inputs and outputs of technological systems are frequently called "flows." The term "flow" is used to indicate passage into or out of the system boundary. Flow helps convey the idea that when carrying out functions, technological systems are dynamic, transforming available inputs into desired outputs.

A flow is not necessarily a non-stop continuous stream of material as in a flow of water from a faucet. Although it is frequently the case that flows are continuous when systems are in operating mode such as the flow of electrical energy into an operating light bulb. More generally, flow designates something that moves across the boundary. Moving a rock into a system, for example, would be considered a "Flow." The rock crossed the boundary although it was a single incident not a continuous process.

In describing technological systems, it is convenient to have some general categories to classify the types of inputs and outputs of these systems. Convenient classifications are materials, energy, and information. These three types of flows are introduced here and then considered in more detail in later sections. Different styles of arrows designate the different flow types in system block diagrams. These are illustrated in Fig. 2.9.

Matter (or materials) forms one class of flows. Matter can be a solid, liquid, or gas. Matter has properties such as mass, color, or shape. Matter is conserved, meaning if matter of a particular mass flows into a system the total amount of mass must either flow out or accumulate in the system. Flow of material can be continuous such as flow of water measured in gallons per minute or refer to a specific fixed amount of mass crossing the system boundary on a one-time basis. In block diagrams of systems, a thick arrow will be used to indicate a flow of material.

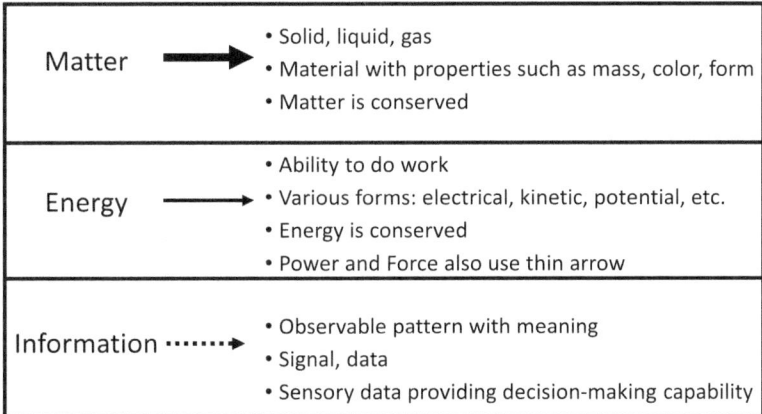

**Fig. 2.9** Overview of three types of flows

Energy, along with related concepts of force and power, constitutes another class of flows. Energy is often defined as the ability to do work. Energy exists in various forms such as kinetic, potential, and chemical. Energy can change from one form to another. The total amount of energy is conserved.

Power is the rate at which energy is transferred. A continuous flow of energy into or out of a system represents power entering or leaving the system. Power is described in units of energy per time.

Force is an influence that can change the motion of an object. Force is frequently described as a push or a pull. Forces result from objects interacting. A system must interact with something external to the system for a net force to be exerted on the system.

Work is defined as force acting over a distance. Work is a type of energy. In this way force and energy are closely associated.

Energy, power, and force are closely related. For sake of simplicity, a thin arrow will be used in system block diagrams to indicate related concepts of energy, power, or force.

Information flows form the third type of input and output. Information can be defined as any observable pattern that can be interpreted and has meaning. In technological systems, information might be a signal or some type of data. Data and signals are often in electrical form. Technological systems process and produce information. Within systems, information frequently serves a control function of system behavior.

Another type of information is sensory data present in the system that a user might experience. In other words, what would a person see or hear from the system such as a video image or audible sound? A dotted arrow indicates a flow of information.

## 2.5     Material

### 2.5.1     Material Defined

What is meant by a flow of material? Material is the most easily identified, or the least ambiguous of the items that undergo a transformation by a technological system. Sometimes the term "Matter" will be used which is synonymous with "Material". Matter has the property of mass. Mass is proportional to the number of protons, neutrons, and electrons present. This is an intuitive concept that mass depends on the amount of material present. Mass is measured in units of kilograms in the SI system and units of pound-mass or slugs in the US customary (English) system.

Mass is conserved. This is an important property of mass from the point of view of technological systems. Material has to go somewhere. In most processes atoms are not destroyed. They can combine and recombine with other atoms. They can have more or less energy, but atoms, and therefore mass, are conserved. The one exception to this is in nuclear reactions which can convert mass and energy. However nuclear reactions are not common in most technological systems and processes.

### 2.5.2     Material Flow Examples

Materials transformation can be an important aspect of some technological systems. Materials transformations constitute a central aspect of technological systems in a wide range of applications. For example, the function of the combine harvester shown in Fig. 2.10 is to harvest and clean grain crops. The harvester is a materials transforming device. The purpose of this technology is to separate the grain or edible part of the plant from the inedible part, the stalk or chaff. The input flow of materials is the entire plant as grown. The mechanism of the harvester separates the grain from the chaff. Two flows of material leave the technological system, one flow of grain and one of chaff. The total amount of mass or material is not changed. The same amount of mass enters the system as leaves the system. However, the leaving mass is separated into grain and chaff.

A manufacturing facility is another example in which the flow and transformation of material is the prime function of the technological system. In the manufacturing facility raw materials enter the system. The purpose of the system is to transform these materials into a finished product. This product leaves the system. The total mass of material is constant. The amount of mass of raw materials is equal to the mass of the finished products plus any waste produced. An example of material flows in manufacturing is depicted in Fig. 2.11.

Materials can undergo chemical reactions within a system. For example, ethanol can be produced from corn or other plant materials. An ethanol plant transforms the material corn into another material ethanol. Figure 2.12 illustrates this process. Other inputs include yeast, water, and chemical enzymes. This transformation will require an input of energy

**Fig. 2.10** Material flow in a combine harvester (iStock.com/dmathies)

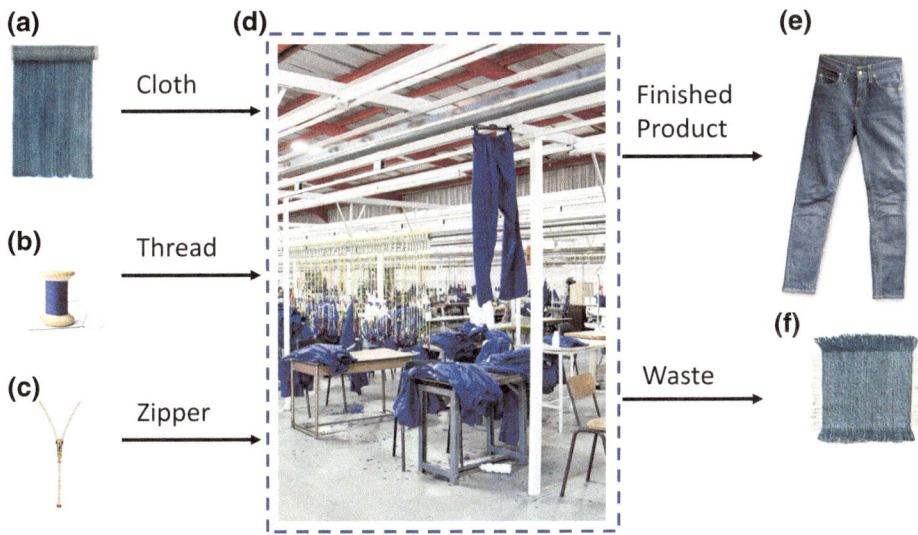

**Fig. 2.11** Material flow in a manufacturing facility: **a** denim cloth (iStock.com/alekleks); **b** thread (iStock.com/LuisPortugal); **c** zipper (iStock.com/urfinguss); **d** textile factory (iStock.com/poco_bw); **e** denim pants (iStock.com/primeimages); **f** denim scrap (iStock.com/inxti)

as well. All of the matter in the corn is not converted into ethanol, so some unconverted grain will also result.

A municipal wastewater treatment plant is an example of a materials transformation process. In this case the system is separating rather than combining materials. The function of the treatment plant is to separate the waste from the water. As shown in Fig. 2.13 the output is clean water and waste separated from each other.

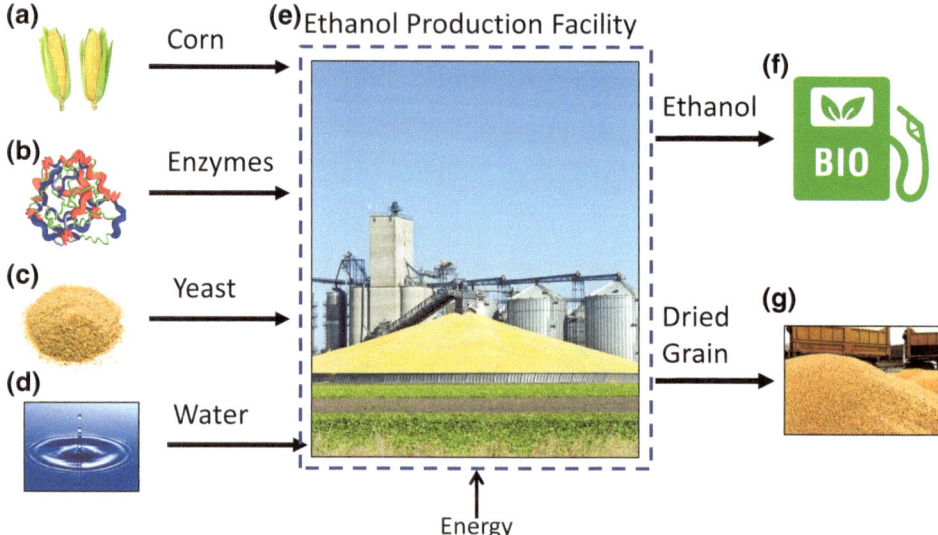

**Fig. 2.12** Ethanol production from corn: **a** corn (iStock.com/fongfong2); **b** enzymes (iStock.com/TimoninaIryna); **c** yeast (iStock.com/KrimKate); **d** water (iStock.com/Paket); **e** ethanol production facility (iStock.com/YinYang); **f** biofuel station (iStock.com/Anastasiia Konko); **g** dried grain (iStock.com/ligora)

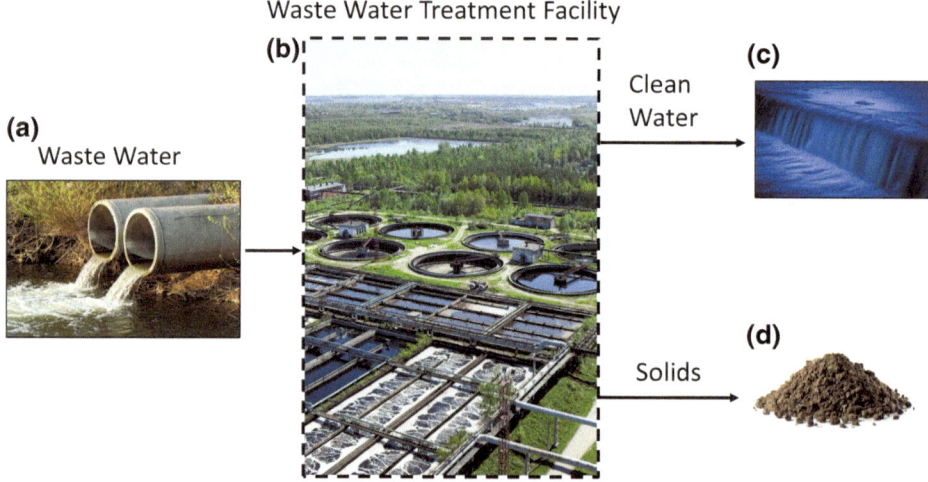

**Fig. 2.13** System that separates material from municipal water: **a** waste water (iStock.com/aquatarkus); **b** waste water treatment facility (iStock.com/antikainen); **c** clean water (iStock.com/Lyudmila Lucienne); **d** solid waste (iStock.com/malerapaso)

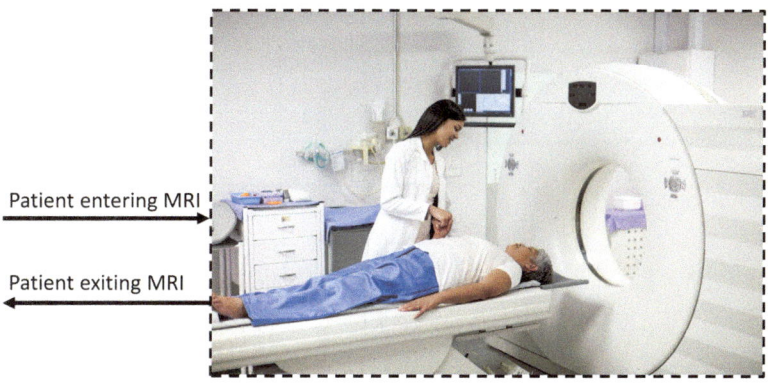

**Fig. 2.14** Material flow into and out of an MRI imaging system (iStock.com/laflor)

It is not necessarily the case that the material moving into and out of the technological system must change form as a result of the system operation. Some circumstances require that a material be transported into a technological system. The same material then leaves the system. For example, in some medical magnetic resonance imaging systems (MRI), the patient must be inside the field of the electromagnet during the imaging process. The system must be designed to transport the patient into and out of the field of the magnet. The function of the MRI is not to create any change in the patient's body but rather to obtain images of internal parts of the body. Material in the form of the patient is moved into and out of the system but no transformation of the patient occurs. This process is depicted in Fig. 2.14.

The conservation of matter creates challenges when a technological system utilizes materials but the function of the system produces materials that are not needed. An example of this is a combustion process such as occurs in an automobile engine. The engine functions to transform the chemical energy available in the fuel and air into mechanical energy of motion which is then transferred from the engine to the driven wheels.

Materials in the form of air and fuel enter the engine. A chemical reaction takes place in the combustion process. In the basic combustion process, the hydrocarbon molecules in the fuel react with oxygen in the air to produce water and carbon dioxide. These products form the exhaust. All of the mass of material that entered the engine is converted into exhaust products. That means as much mass as flowed in must also flow out. Each kilogram of gasoline burned requires about 3.6 kg of oxygen from the air. Consequently, burning one kilogram of fuel produces 4.6 kg of exhaust. In terms of other unit systems, burning one gallon of fuel that weighs about 6 pounds will produce about 26 pounds of exhaust. The conservation of mass is shown in Fig. 2.15.

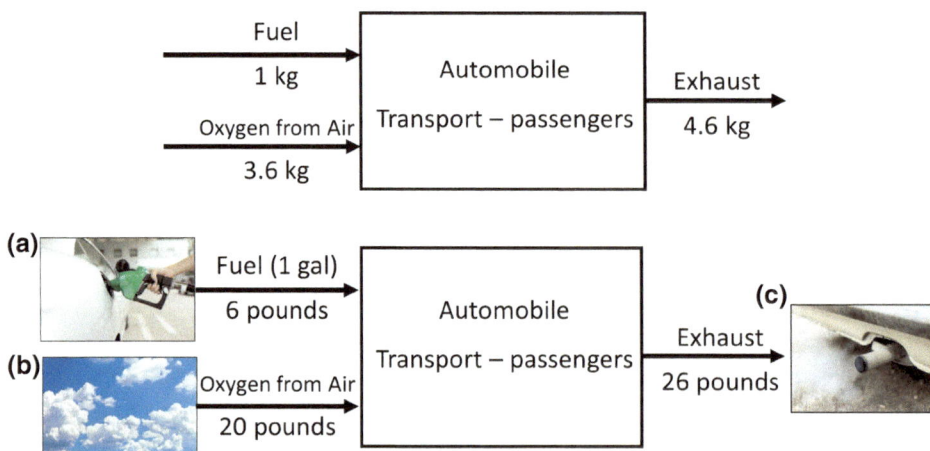

**Fig. 2.15** Fuel and air combine to produce exhaust in an automobile: **a** gas pump (iStock.com/Pho toTalk); **b** sky (iStock.com/xxmmxx); **c** exhaust and tail pipe (iStock.com/Alex_Ishchenko)

## 2.6    Energy, Power, Force

### 2.6.1    Energy

Energy is a vital aspect of many technological systems. The function of most technology requires some type of energy input. In other cases, the entire purpose or function of the technological system is to transform energy from one form to another.

Energy is defined as the ability to do work. Work in turn is defined as a force acting over a distance. Energy then is the ability to apply a force over a distance. Energy exists in many forms such as kinetic, potential, and thermal energy. Energy can change forms; however, the total amount of energy is conserved just as mass is conserved.

A form of energy does not necessarily have to be a force moving over a distance as long as there is the potential or possibility of moving a force over a distance. For example, thermal energy has the ability to do work. Consider the system shown in Fig. 2.16. The weight is sitting on the top of the container. As heat is applied the temperature of the air increases and the air expands. The expanding air raises the weight accomplishing work. Using thermal energy, it is possible to move a force through a distance.

Since energy is defined as a force multiplied by distance, the dimensions of energy are determined by the dimensions of force and distance. In the SI system force has units of newtons and distance has units of meters. So, work and energy have units of newton-meters. In the SI system the units of newton-meters are given the name "joule" or symbol "J."

$$1 \text{ Joule } = 1 \text{ newton } - \text{ meter}$$

**Fig. 2.16** Illustration of thermal energy producing work: **a** kettlebell (iStock.com/jpa1999); **b** lit candle (iStock.com/Rafael Malta)

### 2.6.2  Power

Power is a concept that is related to, but not the same as, energy. Power is defined as the rate at which work is being done or the rate of change in energy. Power can be viewed as an indication of how quickly work is being done or energy being expended. Power is energy divided by the time over which the energy is expended.

$$\text{Power} \; = \; \text{Work/time or Energy/time.}$$

The units of power are Joules per second. This is given the name Watts. One horsepower = 746 W.

An example may clarify the difference between energy and power. Suppose a heavy object like a box is sitting on the floor. Someone lifts the box. A force is being applied over a distance so work is being done equal to the force, in this case the weight of the box, multiplied by the distance over which the force acts. This distance is how high the box is lifted.

Is there a difference between lifting the box quickly compared to lifting it slowly? The work done or energy expended is just equal to the weight multiplied by the distance. The energy needed to lift the box is the same whether the box is lifted quickly or slowly. However, intuitively we know that there must be difference.

The difference between lifting quickly and slowly is described by the concept of power. Power is energy or work divided by the time. So, when the box is lifted quickly the time is less, so the power needed is greater. Lifting the box quickly requires more power, but the same amount of work or energy is needed regardless of the time needed to lift the box.

### 2.6.3  Types of Energy

Energy can take a variety of types in technological systems. In identifying energy, a question that is often helpful to determine is if there is any way to create a force acting through a distance? Table 2.1 lists some common forms of energy that appear in technological systems.

Table 2.1 provides some specific examples of different types of energy. An automobile in motion, the water flowing in a river, and wind all have energy of motion or kinetic energy.

Thermal energy is identified by matter that has a different temperature than its surrounding environment. For example, thermal energy may be identified in a hot cup of coffee. The coffee is at a higher temperature than its surroundings so energy in the form of heat flows from the coffee cup to its surroundings.

Chemical energy is recognized in materials which can undergo a chemical reaction and release energy in the process. This is often thermal or kinetic energy. Examples of materials that may be considered as representing a source of chemical energy might be an explosive, a match, coal, and gasoline. Combustion is not the only reaction that can cause chemical energy to be transformed into thermal or kinetic energy. Iron reacting with oxygen releases heat.

**Table 2.1**  Different types of energy

| Type of energy | Description |
| --- | --- |
| Kinetic energy | Energy due to motion |
| Potential energy | Energy associated with location |
| Thermal energy | Energy due to atomic or molecular motion (heat) |
| Chemical energy | Energy associated with atomic and molecular bonds |
| Sound energy | Energy transport by a mechanical wave of oscillatory compression and displacement |
| Electromagnetic | Energy carried by propagating electromagnetic fields (such as radar, radio waves, microwaves) |
| Electrical | Energy associated with forces on electrically charged particles and the motion of electric charge |

All electromagnetic waves contain energy. Electromagnetic waves include x-rays, ultra-violet light, visible light, infra-red light, microwaves, radar, radio waves, mobile phone, and WIFI. These phenomena differ in the wavelength of the wave. Electromagnetic waves transport energy.

Sound is also a transport of energy. Sound is typically created by something vibrating or moving back and forth quickly. These vibrations can be transferred from one object to another. Since it involves motion sound could be considered as a type of kinetic energy but given its unique nature sound is often considered as a separate category of energy. A speaker receives an input of electrical energy and produces an output of energy in the form of sound. An object producing noise of any kind represents a flow of energy from the object in the form of sound.

Energy can exist in the form of mechanical potential energy. Examples include a stretched rubber band or a compressed spring. In each case the possibility exists to create a force moving through a distance.

More than one form of energy can exist simultaneously. For example, the hot air leaving a hairdryer has both kinetic energy of motion and thermal energy.

### 2.6.4   Energy Transformation

It is possible to change one type of energy into another. In fact, the function of many technological systems is to create a change in energy from one form to another. For example, the light bulb shown in Fig. 2.17 changes electrical energy into energy in the form of electromagnetic waves or light. The function of a wind turbine is to transform the kinetic energy of motion available in the wind into electrical energy. A photovoltaic solar panel converts energy in the form of electromagnetic light waves into electrical energy. An internal combustion engine converts the chemical energy in the fuel into energy of motion and thermal energy.

### 2.6.5   Materials and Energy

Material flows can transport energy into the system. The internal combustion engine demonstrates this case. Fuel and air are materials that enter the system. Simultaneously these materials transport chemical energy. There is both an input of material and energy represented by the same quantity of mass. The same chemical energy does not flow out of the system with the exhaust products. Fuel and air burn and convert chemical energy into thermal energy. The action of the piston in the engine causes the engine crank to rotate. Chemical energy is converted into energy of motion. Energy is transferred out of the system but not simply through the material flow of the exhaust. Some of the chemical energy of the fuel is not converted into kinetic energy but leaves the engine as heat.

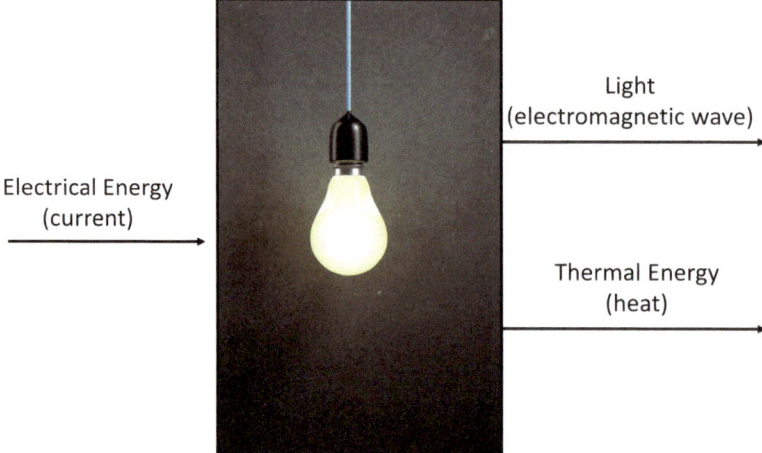

**Fig. 2.17** Incandescent light bulb converting electrical energy into light energy (iStock.com/cha ofann)

Energy can change form but materials transporting energy are conserved. The mass of exhaust products leaving the internal combustion engine is the same as the mass of air and fuel entering the engine. The exhaust does not have the same chemical potential energy as the entering fuel and air, but the mass remains unchanged.

Another example of a material serving a dual role in transporting energy into a technological system is a battery. Many electrical devices utilize batteries. The battery must enter the system at some point. When the battery is used up it is removed and replaced. This represents a flow of materials into and out of the system. However, the battery also carries chemical energy which is converted into electrical energy during operation.

### 2.6.6   Energy Conservation

Whatever quantity of energy that enters a technological system must either flow out or be stored in the system in some fashion. Energy can change form, but it cannot disappear. When a conventional bicycle stops what happens to the kinetic energy of motion? It is converted to heat energy in the brake system of the bike. The heat is eventually transferred to the surroundings. The kinetic energy of motion becomes heat flow to the surroundings.

### 2.6.7   Force

In considering the topic of energy flow into and through technological systems, it is helpful to also address the related subject of force. Force is an intuitive idea. A force is a push or pull and is a phenomenon readily familiar in everyday experience. Force can cause acceleration or in other words a change in the velocity of a mass. In the US customary unit system force is measured in pounds. In the SI system force has units of newtons.

One expression of Newton's second law states that the mass of a body multiplied by its acceleration is equal to the sum of the forces acting on that body.

$$\Sigma F = ma$$

If an object is not accelerating, then the sum of the forces acting on the body is zero. It is critical to notice that this does not mean that zero acceleration equals zero forces, but rather the net result of the forces acting is zero. So multiple forces could be acting on a body with zero acceleration provided the sum of these forces cancel each other in some sense. The direction must be taken into account when evaluating the net sum of forces acting on an object.

The view of forces as being transferred through components can be a helpful perspective in some circumstances. The field of machine component design sometimes employs force flow visualization to depict and analyze how forces are transferred through components. Figure 2.18 conveys the basic approach of force flow visualizations. Force is treated like a fluid that flows through the component.

Force flow visualization differs from a free body diagram used in physics for problems involving the application of Newton's second law. A free body diagram focuses on a particular object or point on the component and shows all of the forces acting on that

### *Force Flow Visualization*
Method for visualizing how forces
are transmitted through components

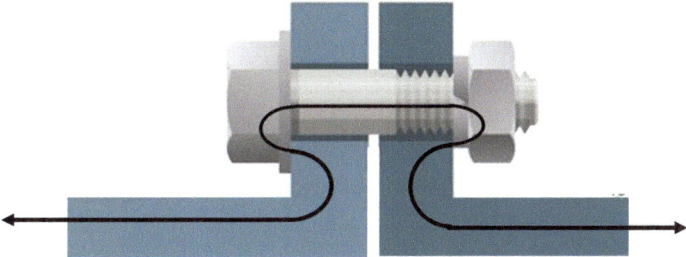

**Fig. 2.18**   Force flow visualization in a flanged connection (illustration by author)

particular point. The forces internal to the object are not considered, but rather forces acting at the object's boundaries as defined in a particular problem.

Force flow visualization can be thought of as a free body diagram made of every part of the object. The forces acting on each piece of the object were shown in a continuous succession. Force flow visualization has also been described as: *"Treat[ing] force like a fluid that flows in and out of the interfaces and through the component.*[3]*"* Similarly, in analyzing the way force is transferred through a system it is suggested to: *"follow the lines of force, approximate paths taken by the force, determined by simple inspection through the various parts, and noting along the way any sections suspected of being critical.*[4]*"*

Figure 2.18 shows an object made of two components fastened in a bolted connection. Consider that a force is applied to the left side of the component. Assume that the object is stationary. If that is the case, then the force must be counteracted by a force of equivalent magnitude acting to the right. But what happens inside the object? Force flow visualization illustrates this. The force must be transferred from the slender left part to the flange. From the flange the force is transferred to the top of the bolt holding the components together. The force then flows through the body of the bolt to the other side. It is then transferred to the opposite flange and finally to the other end of the component. This process demonstrates how the force applied at one end is transmitted through the object.

In this book, diagrams illustrating the interactions between systems and components will use a thin-lined arrow to represent energy, power, or force. The precise nature will be clarified through text labeling. The single arrow type for the related, but not identical, concepts of energy, power, and force reduces diagram clutter in most instances.

## 2.7  Information

### 2.7.1  Defining Information

Information is a major category of inputs and outputs of many technological systems. Information is a broadly used descriptor with numerous context-specific interpretations. Information as "data" or a "signal" are frequently used terms when describing a class of interactions of technological systems with their environment, other technological systems, and the interactions of components within a system.

Information is any observable pattern that has meaning and can be interpreted. Information or data can be discrete such as a sequence of characters or numbers. This is sometimes referred to as digital information especially when the sequence is represented by the binary characters of 0 and 1. The information pattern can also be continuously varying rather than discrete. For example, the variation of air temperature over a 24 h time

---

[3] Ullman (1994).

[4] Robert C. Juvinall and Kurt M. Marshek, *Fundamentals of Machine Component Design*, J. Wiley, (1991) p. 52.

period or the varying intensity of sound from a musical instrument are not restricted to discrete changes but can alter over a range. In contrast to digital, this type of information is referred to as analog information.

Some technological systems can be viewed as primarily information processing or transforming systems. Figure 2.19 shows a mobile phone. The system receives information input in the form of a radio signal. The system transforms this input information into outputs of audio and video information that are perceived by the user of the device.

Figure 2.20 is another example of a technological system that transforms information. The system is a DVD player. The input is the DVD that contains encoded video and audio information. The DVD player transforms the information encoded on the DVD into an output audio and video signal that can be displayed on a screen. Similarly to the case of fuel transporting chemical energy into an internal combustion engine, the DVD, a material, can be viewed as transporting information into the system.

Information can serve a control function in technological systems. A TV and a remote are seen in Fig. 2.21. The remote output is a control signal input for the TV. The user makes a selection and presses the appropriate buttons on the remote. The remote produces a signal that contains the user choice information. The control signal is received by the TV and the appropriate system response such as volume increase, mute, or channel change occurs.

In technological systems information is being transferred from one place to another. Information is transferred or moves into the system from the outside or the system transfers information past its boundary to something external to the system. Within the system, some component interactions can involve transfer of information from one component to another.

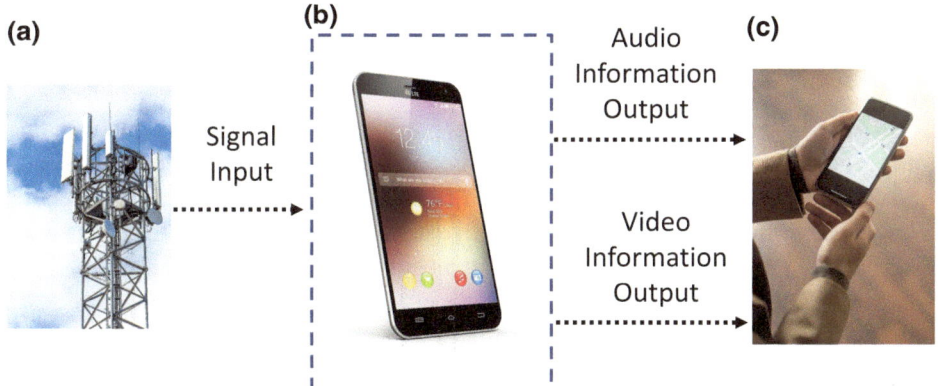

**Fig. 2.19** Mobile phone information input and output: **a** cell tower (iStock.com/MartinFredy); **b** mobile phone (iStock.com/scanrail); **c** mobile phone with user (iStock.com/shironosov)

**(a)**                    **(b)**                                              **(c)**
DVD with                DVD Player                                        Audio Visual
Information                                                              Display Screen

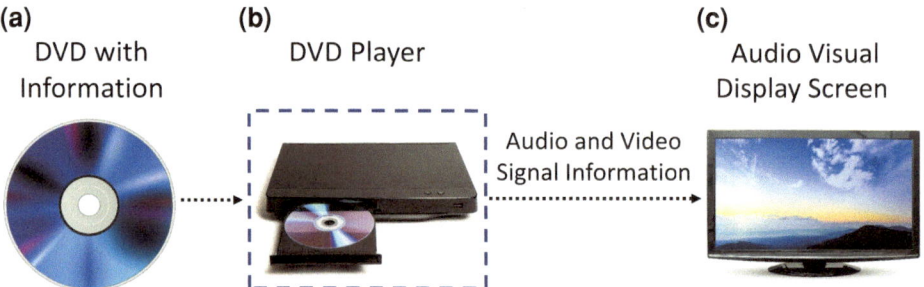

Audio and Video
Signal Information

**Fig. 2.20** DVD information input and output (iStock.com/Cobalt88): **a** DVD (iStock.com/GetUpS tudio); **b** DVD player (iStock.com/); **c** audio visual display (iStock.com/Jay_Zynism)

**(a)**                                                    **(b)**

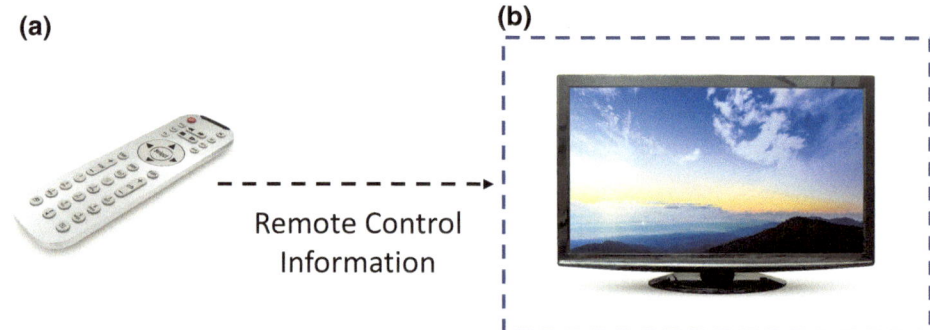

Remote Control
Information

**Fig. 2.21** TV remote control information flow: **a** remote control (iStock.com/Aleksandr_Petrun ovskyi); **b** audio visual screen (iStock.com/Jay_Zynism)

### 2.7.2   Information Transfer

The transfer of information between elements can also be described as communication of information. Communication of information occurs across the system boundary. Components also communicate information within the system. A framework for characterizing the communication of information originally advocated by Shannon (1948) is helpful in describing the flow of information within and between technological systems.

Shannon conceptualized a general problem of communication as outlined in Fig. 2.22. The information starts as a "message" at the source to be communicated to a destination. For this to happen the message must be encoded by some type of "transmitter" into a form that is compatible with being transferred along a communication "channel." The transformed message is called a "signal." The signal on the channel becomes an input to a "receiver" that decodes or transforms the signal into a form of the message that can be

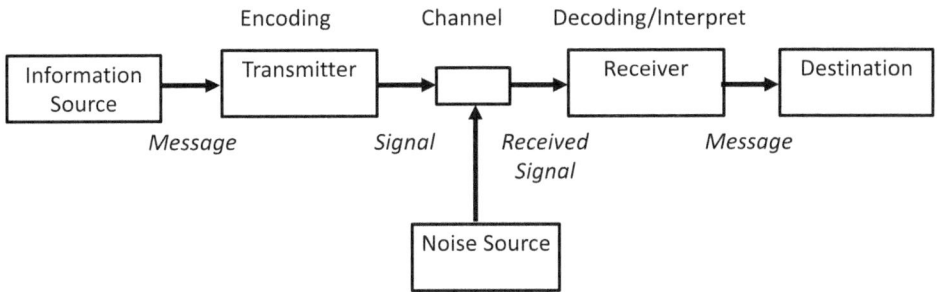

**Fig. 2.22**  Shannon's Model for Communication of Information

interpreted at the destination. The possibility of "noise" or some type of degradation of the signal as it is being transferred along the communication channel is also considered.

Shannon's theory of communication of information applies broadly to all types of modern technological communication systems but was originally developed in the context of twentieth century telegraph, telephone, radio, and television. Telegraph communication provides an example of the basic features. The "message" is a statement written in English using the standard alphabet. For transmission, the message is encoded into Morse code in which each letter is assigned a unique series of long and short pulses (dots and dashes). In this form the long and short pulses are sent by the transmitter along the channel in the form of voltage pulses along a wire. At the receiver, the long and short pulses are converted into long and short audible clicks. These are read or "decoded" by the operating receiver into their corresponding alphabet characters. The transmitted message or the information is received and "read" by the destination.

### 2.7.3  Quantifying Information

Information can be quantified similarly to energy and material. Quantifying information is readily described when considering information in discrete form that can be represented by a series of 0 and 1 of binary digits. A binary digit, or the presence of a 0 or 1, is called a "bit." The number of bits needed to represent a specific message can be considered as the size or the amount of information, some number $N$ of bits.

Considering the communication of the message along the channel from the transmitter to the receiver provides context for the rate of information flow. The number of bits sent and received in a particular time period is one measure of the information rate. For example, $N$ bits per second (abbreviated $N$ bps) is a measure of the rate or speed of information flow in digital communication.

### 2.7.4    Information, Material, and Energy Distinctions

According to physics, the universe consists of matter and energy therefore, strictly speaking, information is either matter or energy. However, the unique role of information in technological systems necessitates separate treatment. Information is an observable pattern that can be interpreted and has meaning. The pattern must occur in some medium. That medium can be in the form of energy or material.

Energy is often a medium for information. This can take forms such as sound, light, radio waves, or electric current. Typically, the magnitude of the energy is usually small for information flows compared to other energy flows in the system. Energy is identified by its ability to do useful work. Information is not considered significant because it can be converted into useful work; it is significant because it contains an interpretable pattern that is significant either for interactions within the system or for interactions between the system and external entities.

The observable pattern of information can also be associated with a material. Material can be the medium in which the information is embedded. Material has physical properties such as mass and volume, but when material is serving as a medium for information, the presence of the pattern is a primary interest. Examples of material serving as the form containing information include a DVD, a barcode, a written message on a piece of paper, a photograph, and a magnetized tape. When information is contained in a material the system must also accommodate the physical aspects of the information medium.

### 2.7.5    Information Non-Conservation

Matter and energy are conserved quantities. Conservation of information is not usually thought of as a conserved quantity. A DVD contains information. If the DVD is shredded the pattern containing information is lost however the total mass of the shredded pieces equals the mass of the original DVD. The mass of the DVD is conserved but the original information content is lost.

A mobile phone user sends a text message. Information leaves the phone via a signal. The information content of the phone has NOT necessarily decreased in this situation.

### 2.7.6    Examples of Technological Systems Emphasizing Information

The examples of technological systems emphasizing information shown in Figs. 2.19, 2.20, 2.21 and 2.22 can be described more precisely using the vocabulary of communication theory. For the mobile phone shown in Fig. 2.19, the message to be received is the eventual video and audio output as perceived by the user. The message is first encoded on a radio wave and transmitted by the cellphone antenna tower. The radio wave serves as the channel. The radio signal is received by the mobile phone. The phone extracts the information from the radio wave and produces the output audio and video for end-destination user.

In the example of the DVD player in Fig. 2.20, the encoding is the information on the DVD. The DVD itself is the channel that transports this information into the system of the DVD player. The player is the receiver that extracts the audio and video signal from the DVD and sends this information to the TV for display.

For a TV remote, as shown in Fig. 2.21, the information source is the user and the message is the user's desired control action of the TV. The user presses buttons indicating their choice. The remote encodes this message as a signal consisting of pulses of infrared light emitted from the end of the remote. The channel here is the light that travels from the remote to a sensor on the TV. The TV converts the received signal into the appropriate control action and implements the user's choice on the TV.

Some technological systems have a primary function of producing some type of information. A smoke detector is one example. The function of this system is to communicate information that smoke is present to people near the device. The major inputs to the system are smoke and electrical energy. When smoke is present the smoke detector produces a characteristic loud blaring sound conveying information that smoke is present.

Figure 2.23 illustrates the inputs and outputs of a smoke detector. The message is the presence or absence of smoke. When enough smoke enters the device, the internal components encode this message as a signal that can be conveyed on a channel, in this case a sound wave. The ear of a nearby person is the receiver that decodes the sound wave signal into a message interpretable by the brain of the person that is the destination of this information.

A digital thermometer is another example. The thermometer shown in Fig. 2.24 is an information-producing device. The function of the thermometer is to measure and display information about temperature. A major input is energy in the form of heat from

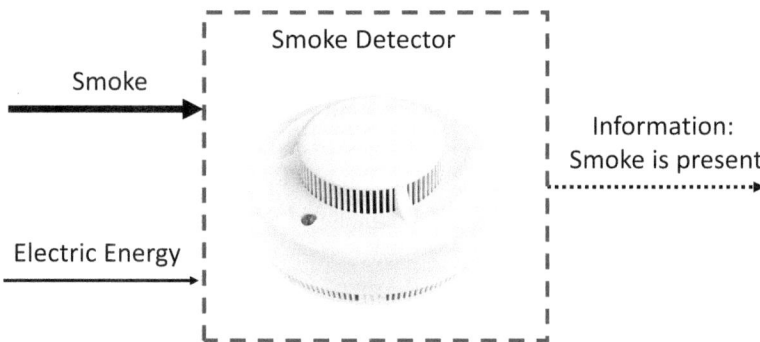

**Fig. 2.23** Smoke detector produces information about presence of smoke (iStock.com/IgorKoval chuk)

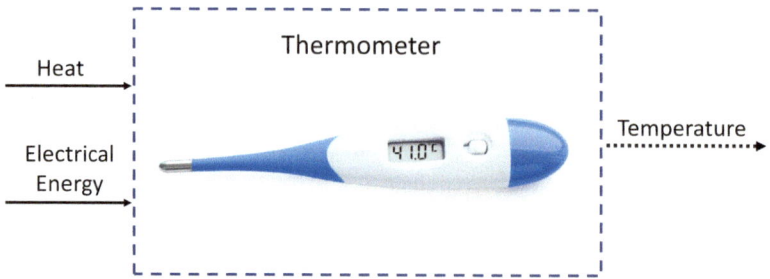

**Fig. 2.24** Primary output of a thermometer is information about temperature (iStock.com/deepbl ue4you)

the object being measured. The thermometer's internal components also require electrical energy provided by a battery. The major output is information about temperature.

The thermometer can be viewed as encoding the amount of thermal energy present as a temperature. The signal is the numerical display of a number. The user sees the display and interprets the number.

Some technological systems involve a user or operator carrying out control functions. The user requires information about the condition of the system. Information leaves the system that helps the user to determine the status of the system. Some of these information flows may be produced deliberately by system components while others may be information produced by incidental physical effects occurring as part of system operation.

Consider a functioning coffeemaker as shown in Fig. 2.25. Some system information can be identified by asking questions such as: "How does the user know that the system is carrying out its function?" The coffee maker brews coffee. The inputs are water, coffee grounds, a filter, and electric current. The output of the system is hot brewed coffee. To identify potential flows of information from the system we can ask the question: how would the user know the coffeemaker is working? Or what indicates that the coffeemaker is operating as expected? What sources of data describe the status or condition of the coffeemaker?

Information leaving the coffee maker system would include: the smell of coffee, the level of coffee in the carafe, the color of the coffee, the temperature of the coffee, and the status of the ON/OFF light of the coffeemaker. All of these items provide data or information about the condition of the system. Figure 2.25 illustrates information describing the coffeemaker system. For clarity, the material and energy inputs and outputs have been omitted.

Coffee smell

Level of coffee in carafe

Coffee color

Temperature of coffee

Light from ON light

**Fig. 2.25**  Coffeemaker as an example of information produced by a system (iStock.com/benimage)

## 2.8    Flows with More Than One Quantity of Interest

In some instances, a major input or output of a system will convey more than one quantity of interest. These may interact separately with other parts of the system. It is helpful to have a diagrammatic protocol for representing situations in which one of the inputs or outputs has more than one property of importance to describing the system behavior.

For example, a hydroelectric turbine-generator is used to produce electrical energy from flowing water. The input is water in motion. The turbine and generator convert the energy of motion into electrical energy. Electrical energy exits the system via transmission lines into the electrical energy distribution grid. The water leaves the system.

In representing this system, a goal is to convey the process visually in a direct and clear manner. The central feature is the transformation of energy of motion into electrical energy. Figure 2.26 shows a block diagram representing the transformation. In Fig. 2.26, the input shows water and its associated energy of motion. Two separate arrows are used but are shown as linked with vertical lines. The water is transporting the energy into the system as energy of motion. This one flow has two properties of interest: mass flow and the energy of motion. Within the system this energy of motion is transformed into electrical energy. Electrical energy leaves the system. The water also exits.

Connected parallel lines with a line linking to the two arrows indicate that they represent one physical flow but with more than one quantity of interest. An alternative would be an arrow made of multiple parallel lines representing the quantities of interest but with a single arrowhead as shown in Fig. 2.27. The first approach of separate linked arrows is easier to produce using computer-based diagram-making applications and will be the alternative used here.

Hydroelectric Turbine Generator

Energy of Motion

Electrical Energy

Water

Water

Convert energy of motion to electrical energy

**Fig. 2.26**  Hydroelectric turbine (iStock.com/YinYang)

**Fig. 2.27**  Equivalent
approaches for designating a
flow with multiple properties
of interest

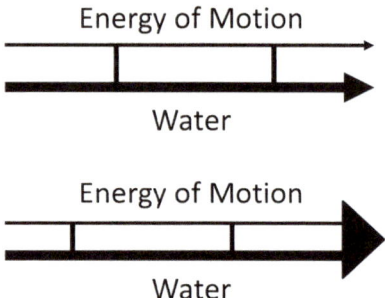

Energy of Motion

Water

Energy of Motion

Water

Materials are conserved in the turbine system. The same amount of water that enters also leaves the system. Energy is conserved. The system extracts energy of motion from the water and converts this into electrical energy. The use of the double parallel arrows linked together for incoming water and energy of motion helps to make the specific transformation of inputs into outputs more visually apparent. The turbine system extracts the energy of motion from the water and converts it into electrical energy.

The water leaving the system still has some energy of motion. Otherwise, it would not be moving at all. However, the energy of motion is substantially reduced. It is simply water leaving the system. The exiting mass flow of water is represented by a single wide arrow used for material flows.

## 2.9    Phases and Modes of Operation

The function of a technological system can be described as causing changes in the flows of material, energy, and information through the system. One aspect to emphasize is the flows of materials, energy, and information through a technological system may not be the same at all times during system operation. In fact, it is more often the case that most complex products will have different phases of operation or different operating modes in which the flows of materials, energy, and information are different.

Consider a smartphone as an illustrative example. Like many consumer products it is possible to identify multiple modes of operation with different inputs and outputs. In one mode the phone might be only recharging the battery. The battery is being charged so electrical energy would flow into the system. The energy is stored in the form of chemical energy in the battery. There would be no sound or video signal leaving the device. There might be a flow of information that indicates charging status.

In another mode of operation, the user of the phone might be sending and receiving text messages. Over this period of time, there would be information in the form of text messages encoded in radio waves entering and leaving the system. Information from the visual display leaves the system to be observed by the user. The system uses electrical energy to accomplish all of these operations but no electrical energy enters the system. The electrical energy used is stored energy from the battery.

Using the smartphone to take and send a photograph is another example of a different mode of operation with a different set of system inputs and outputs. Let the time period under consideration extend from the user taking and subsequently sending the photograph. Light information from the scene being photographed enters the system. In the system, the lens and internal components encode this visual information as an image file stored in the system. When the user sends the image to a contact, the image file is encoded into a format compatible with the outgoing signal that leaves the system. The system again uses electrical energy previously stored in the battery.

These examples of some smartphone activities illustrates that in describing behavior of some technological systems there is a need to specify the mode of operation and time period over which a system is observed.

## 2.10  Overall Function: The System-Surroundings Interaction

### 2.10.1  Describing the System-Surroundings Interactions

It is helpful to have an approach for describing the interactions between a technological system and its surrounding environment. A goal is to identify and describe the major transformations that the system accomplishes in the physical world. Essentially, this involves identifying the inputs needed to bring about these transformations and specifying the outputs produced. This can be a useful framework for understanding technological systems. The overall function perspective helps to reduce the distraction of detail and condenses those details into a focus on essential features of the system operation.

It is well-accepted that the complexity and nuanced behavior of the physical world exceeds the degree of detail that can be included in any one type of model or representation. However, representations with strategic simplifications can be useful in communicating significant features and providing useful insights. Technological systems are just one type of system. Other systems may benefit from different types of analyses. Also, there is no one unique approach to applying a systems perspective on any given situation. Describing the interaction of a technological system with its environment can provide useful insights, but it is not claimed to be absolute or exclusive of other system analysis methods.

The overall system function view describes a technological system in an engineering framework. The overall system function provides a starting point for more detailed analyses. This viewpoint is generic and can be applied to any technological system regardless of simplicity or complexity. Establishing a proper starting point is essential. It is not possible to analyze internal workings of a complex technological system unless the inputs and outputs are well-defined and it is clear what the system is supposed to do.

An approach for describing the interactions between a technological system and its surrounding environment is a potentially iterative process of the following steps:

1. Establish the system boundary.
2. Identify overall system function (Verb-Noun).
3. Determine phase or modes of operation to be described.
4. Identify major inputs and outputs.
5. Classify inputs and outputs in terms of material, energy, or information.

1. *Identify the system boundary.* Specify the boundary between the system and the surrounding environment. What constitutes the technological system and what is outside the system should be clear. The environment of a particular technological system is not only the natural environment but also any other systems which may interact with the system in question. The system boundary need not correspond to the specific physical contours of the system. The boundary can be envisioned as an imaginary box enclosing the system.

2. *Specify the overall system functions (Verb-Noun).* A next step is to clearly describe the problem solved by a particular technological system. This is the overall system function. It is helpful to express this in verb-noun form. The overall system function illustrates that causing change is an inherent attribute of technological systems. Creating some type of transformation of provided inputs into desired outputs is a fundamental nature of technological systems. Solving problems by causing change is an essential property of technology.

3. *Describe the operating mode or phase to be described.* The modes or phases of operation that are being considered are specified. Technological systems can have different aspects of operation which are defined by different types of inputs and outputs. In

some cases, iteration between definition of operating mode and the system function might be needed.

4. *Identify the major inputs and outputs.* The system receives inputs from the surrounding environment across the system boundary and provides certain outputs. Some inputs and outputs are less obvious or of a smaller magnitude than others. As in development of any model or description some determination of priority of significance will be needed.

   To help describe the system function and identify inputs and outputs it can be helpful to ask the question: "What is happening, what is observed when the system operates?" Sometimes imagining or visualizing operating the device can be useful in identifying inputs and outputs. Alternatively, imagine watching a video of the system in operation. What moves into or out of the scene? These things are potentially inputs and outputs of the system. The inputs and outputs need not just be physical objects. Inputs and outputs can be things like noise, heat, electric current, or light.

   Considering the use of the technological system as an interaction between the user and the system facilitates the analysis. What are the required inputs or what does the system expect from the user? What outputs cross the system boundary? What does the user expect from the technology?

5. *Classify inputs and outputs in terms of material, energy/power/force, or information.* For technological systems it can be helpful to classify the inputs and outputs in terms of material, energy (or related concepts of power and force) or information. For many technological systems, energy in some form must be an input. This might appear in the form of electric current, supplied by batteries or an electrical cord.

Information might appear as a type of input in the form of control information. This may be carried into the technological system through interfaces such as push buttons or a computer keyboard. Some type of information output is often a specific desired output of a technological system such as audio or video information. Information output is also common as deliberate or incidental cues to allow the user to discern system status.

Describing system function in this way illustrates the most important and essential features of system operation. Subsequent and more detailed analyses will be incorrect if the starting point is incomplete or incorrect. The system function describes the problem solved or need fulfilled by the particular technology. The overall system function describes what the technological system does—not the internal workings or how the transformation of inputs into outputs is accomplished. Once the system "what" is made clear, then the internal "how" can be examined.

## 2.10.2 Microwave Oven Example

As an example of creating an overall system function model of a technological system, consider the microwave oven shown in Fig. 2.28.

1. *System Boundary* In this case, the identification of the technological system as distinct from the surrounding environment is straightforward. The microwave unit itself is considered to be the technological system. Envision an imaginary box surrounding the microwave designating the boundary between the system and the outside world. The system boundary indicated by the dotted line is drawn around the microwave oven. This establishes that the microwave is the technological system under consideration and everything else constitutes the external environment in which this system operates and interacts. Notice that the electric cord used to plug in the microwave is "cut" by the system boundary.
2. *System Function* A microwave oven is usually used to cook or heat food. The verb-noun combination: "heat-food" describes the function of the microwave in verb-noun form. Possible synonyms include "cook-food," or "heat-water." Expressing the function as "heat-water" acknowledges that the microwave heats or cooks food by heating the water contained in the item to be heated.
3. *Phase of Operation* Assume the microwave is functional, accessible to the user, and plugged into an active electrical power source. The most common sequence of operations is a user puts something to be heated in the microwave, turns it on, and retrieves the item at the completion of the heating cycle. The time period under consideration starts with the user intending to heat food in the microwave and ends when the heated food is taken away by the user.
4. *Inputs and Outputs* The goal is to describe the major interactions between the microwave and its surrounding environment. Envisioning the microwave oven going through a sequence of normal operations heating food, what items cross the boundary into the system? Food along with its container or packaging or whatever item is to be heated crosses the system boundary into the system. The hand of the person operating

**Fig. 2.28** Microwave oven system boundary (iStock.com/scanrail)

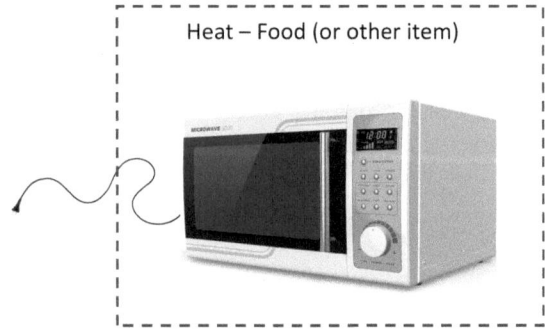

Heat – Food (or other item)

or using the oven crosses the boundary multiple times during typical operation. For example, the user opens the door, puts the food in the oven, closes the door, and sets the controls. The hand is bringing material into the system and also providing control information. During operation, electric current enters the system. This can be recognized by recalling that the system boundary intersects with the electrical cord of the microwave.

What outputs occur during system operation? The heated food or other heated item exits the system. In addition to these items, other outputs include the sounds made by the microwave during use. These sounds comprise the noises from the door opening and closing, sounds from fans running during operation, and any beeps or other noises produced by the microwave as indicator signals to the operator. Light also leaves the system as produced by the clock, timer, or other indicators.

The inputs and output flows of the microwave oven shown in Fig. 2.29 are an initial simplified block diagram of the system using labeled arrows to indicate the major inputs and outputs during the operating sequence analyzed.

Classification of flows in and out as materials, energy, and information provide additional insights into characterizing the interaction of the microwave technological system with its environment. Figure 2.30 depicts the results of this analysis.

On the input side the food to be heated is a material. The electric current represents an input of energy. The hand is a source of control information for the microwave technological system. This is somewhat of a subtle issue but illustrates a central aspect of many technological systems, namely control. The user's hand enters the system, pushes various buttons and the microwave heating starts. The hand is a material and pushing the buttons involves applying a force, but what is most significant about the interaction of this system with its environment is this action is carrying out a control function for the microwave.

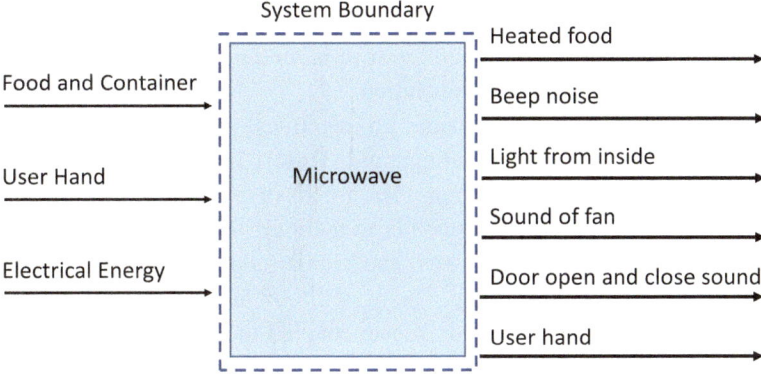

**Fig. 2.29** Inputs and output flows of the microwave oven

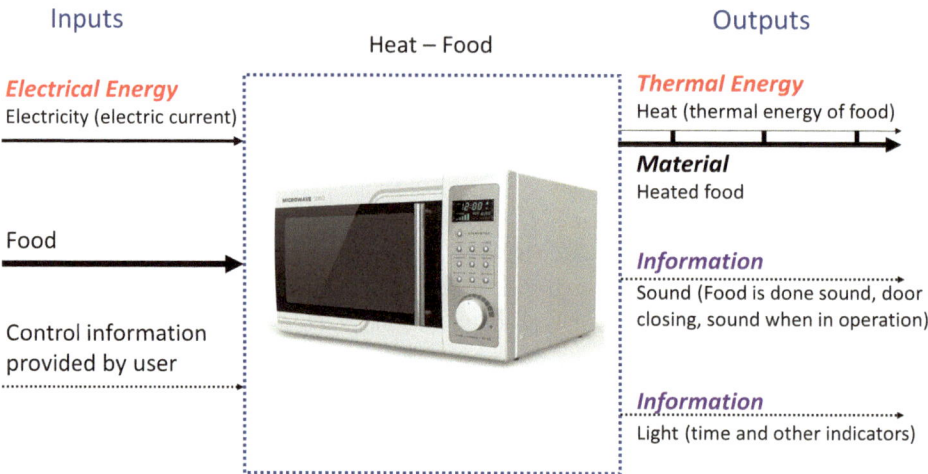

**Fig. 2.30** Overall system description of microwave oven: microwave oven (iStock.com/scanrail)

Selecting the amount of time to "run" and hitting the "start" button are providing control information to the microwave system.

The food being transported by the user's hand into the microwave could be depicted as two materials linked together. The hand plus the food. Or it could be depicted as a single arrow labeled as "hand and food." In this instance, the arrow was simply labeled as food. The decision being that "hand and food" complicated more than it clarified and a single material arrow labeled food was most straightforward and clear.

Information also leaves the system. For example, various sounds leaving the system can be interpreted for their information content. The door makes a characteristic sound when it closes. This provides information to the operator that the door is properly closed. The microwave may have an internal fan that makes noise when it is operating. These noises are information to the user that the microwave is operating. Most microwaves have an internal light that is on when the microwave is running. This light is also information to the user that the microwave is in operation.

The design of the microwave includes an intentional sound such as a beep noise that is produced at the end of the operating time. This is information to the user that the microwave has completed its operation. The microwave may include a time display or other text display that provides information to the user.

Considering the outputs of the system, the heated food is material. Energy is a system output in the form of thermal energy. The heated food has thermal energy. In terms of a visual indication the thermal energy is shown coupled to the food.

The requirements for conservation of energy and conservation of materials are preserved in the process analyzed. Energy enters as electrical energy and is transformed into heat energy that leaves with the hot food. Other uses of energy in the system of much smaller magnitude are not explicitly included. For example, some electrical energy goes to powering the internal platter that rotates the food. Electrical energy is used by the internal light. These much smaller energy uses are not included for the sake of clarity in depicting the most significant energy transformations in the system.

Material is also conserved. Other than an unintentional food spill or splatter, no material stays in the system. The food enters the system and the same amount of food exits. The mass is unchanged.

The overall system function facilitates a succinct description of the essential transformations accomplished by this technological system. The microwave oven is an energy transformation system that receives electrical energy input and transfers this energy to the food in the form of heat. The major inputs of food and electrical energy become an output of heated food. The user provides control information to the system and receives control information signals leaving the system such as the beep noise indicating the cycle is completed.

In summary, it is seen that the overall system function framework can be applied to this appliance providing a description of the problem solved, the inputs required, and the outputs produced. The operating mechanism within the microwave oven itself remains unspecified at this point. The next level of system description will look within the system to identify the internal components and their interactions. Analysis within the system depends upon correct inputs and outputs at the system boundary.

## 2.11 Engineering Fields and Major Functional Transformations

Consideration of the major transformations of system inputs into outputs illustrates the differences between some of the major fields of engineering. Figure 2.28 shows an overview of the primary activity that goes on in four major engineering fields. Engineering fields tend to specialize in producing technological systems that transform particular types of inputs and outputs. This characterization is not exclusive, but gives a general character of each field.

Chemical engineering develops technological systems that transform flows of materials. Civil structural engineering creates systems to direct forces. Electrical engineering produces systems to manipulate and transform electrical energy and information in electrical form. Mechanical engineering often develops systems to transfer mechanical energy or transform one type of energy into another.

A focus of chemical engineering is to transform flows of material matter. The transformation usually involves chemical reactions between some of the input materials. For example, an oil refinery is an application of chemical engineering. Crude oil and energy are input. Through a series of processes the crude oil is converted into products such as gasoline, kerosene, LP gas, and diesel fuel.

Another example of chemical engineering is the production of ABS plastic. ABS plastic is used in a variety of applications ranging from automobile parts to toys. The chemical engineering process utilizes the input materials which include propylene, ammonia, butadiene, ethylene and benzene. Through a series of reactions and intermediate products the final ABS plastic is produced.

Civil structural engineering can be viewed as transforming or directing flows of forces. A bridge is a typical product of civil engineering. One way to view the function of the bridge is to transfer the load resulting from the traffic through the structure to the earth. The function of the bridge is to support the weight of the traffic by transferring the forces to the earth.

Electrical engineering develops technological systems with an emphasis on devices that transform flows of electrical energy and information represented by electrical phenomena. A mobile phone is a typical example. The inputs are the radio signal and electrical energy from the battery. The basic function of the device is to convert the radio signal to sound and visual display information.

Mechanical engineering is a broad discipline but much of mechanical engineering is concerned with changing energy from one form to another. Consider three products that fall under the umbrella of mechanical engineering: an internal combustion engine, a refrigerator, and a solar hot-water heater. The internal combustion engine converts the chemical energy available in the fuel and air into the kinetic energy of motion. The refrigerator utilizes the kinetic energy of motion from an electric motor to create a flow of heat energy from the refrigerator, thus reducing the temperature inside the device. The solar hot water heater converts the energy of the light, a type of electromagnetic wave, into the thermal energy of the hot water.

Mechanical engineering also includes systems that transform flows of mechanical energy (sometimes referred to as kinetic energy). A bicycle for example is a mechanical engineering system. The energy input from the rider is transformed into kinetic energy output of the bicycle wheels.

While each discipline of engineering addresses a particular category of technology, most consumer products are complex technological systems that involve the work of a variety of engineering fields. Consider an automobile. While the basic aspect of the engine may be the domain of mechanical engineering, the automobile also utilizes electronic components in its operation. In addition, the materials such as plastics, paint, glass, and rubber that are products of chemical engineering. The roadways upon which the automobile travels are developed through civil engineering (Fig. 2.31).

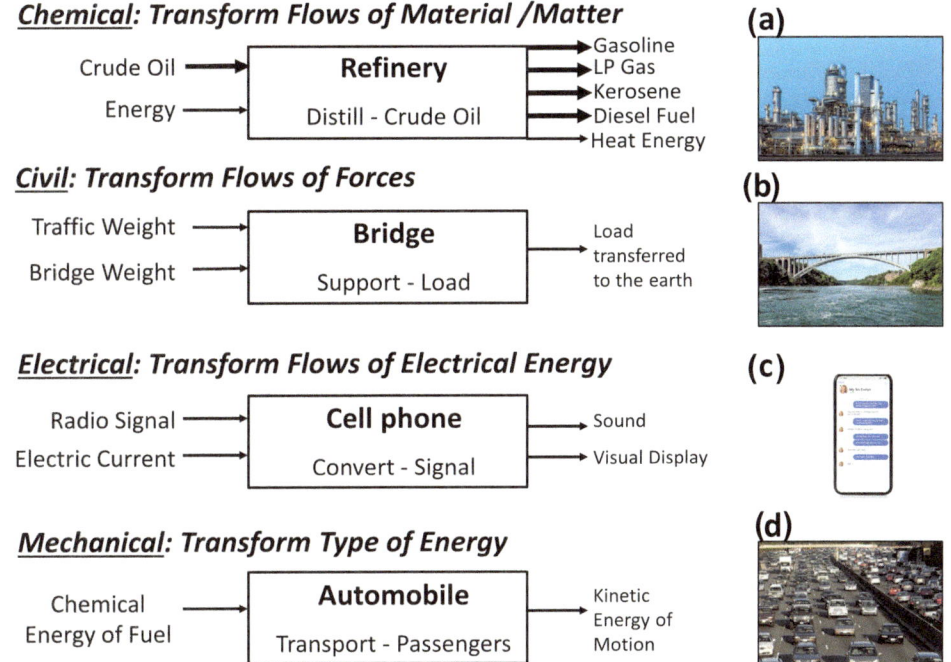

**Fig. 2.31**  Transformations associated with major engineering fields: **a** oil refinery (iStock.com/Bim); **b** bridge (iStock.com/Askolds); **c** mobile phone (iStock.com/grinvalds); **d** automobiles (iStock.com/gdigital)

## 2.12   System Efficiency

The interaction of the system with its environment identifies the inputs and outputs. This type of analysis can be useful in determining system efficiency. A generic definition of efficiency $\eta$ can be defined as a ratio of output to input.

$$\eta = \frac{Output}{Input}$$

The system inputs and outputs are typically the main concern of the end user. The user provides the inputs in some manner and benefits from the outputs. In many technological systems some subset of one or more of the inputs represents an expense or is in some way limited. It is generally desirable then to obtain as much output as possible for the limited or costly available input. System efficiency can be a helpful characterization of the effectiveness of the technology in producing what is desired from what is available.

Photovoltaic Solar Cell

Solar Light Energy 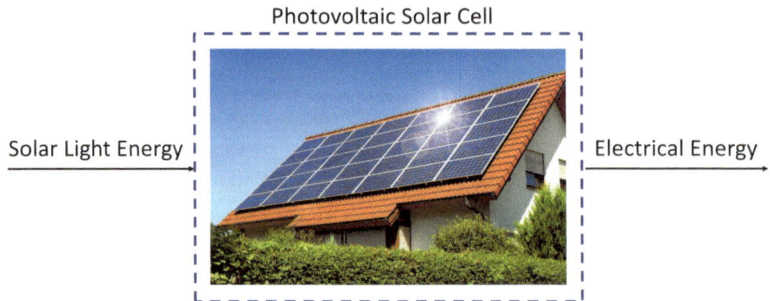 Electrical Energy

**Fig. 2.32**   Photovoltaic solar cell (iStock.com/Smileus)

Due to irreversibility such as friction or other limitations on the processes occurring within the system, efficiency is less than 100%. However, it is usually desirable for the system efficiency to be as high as possible.

Photovoltaic solar cells offer a typical example of efficiency. The function of photovoltaic cells is to convert solar energy into electrical energy. The light from the sun is the input. Users of photovoltaics would like as much electrical energy as possible from the available solar energy (Fig. 2.32).

$$\eta = \frac{\text{Electrical Energy output}}{\text{Available Solar Energy input}}$$

A common photovoltaic cell typically will have an efficiency between 18 and 22%.

A gas furnace is used to heat homes in cold climates. The interaction of this system with its surrounding environment is shown in Fig. 2.33. The function of the furnace is to heat a flow of water and then the hot water is sent to radiators to heat the rest of the house.

The chemical energy available in the natural gas is converted into thermal energy by the combustion process in the furnace. The transformation of major interest to the end user is the amount of heat transferred to the water. Some heat is lost due to the heat energy exiting with the hot exhaust. In this case the efficiency is the useful energy transferred to the water compared to the available energy in the natural gas.

$$\eta = \frac{\text{Thermal energy transfered to water}}{\text{Available energy in the natural gas}}$$

A typical home gas furnace may have an efficiency between 75 and 95%.

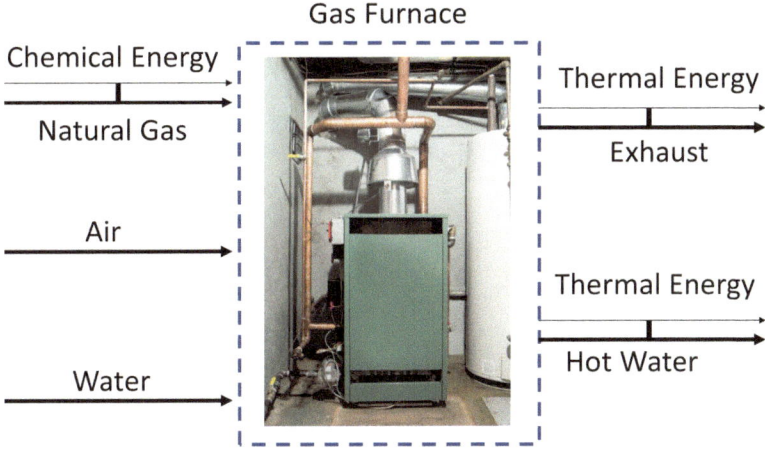

**Fig. 2.33** Gas furnace (iStock.com/nycshooter)

## Bibliography

Dym, Clive L., and Patrick Little. *Engineering Design: A Project-Based Introduction.* Wiley, 2004.

Floridi, Luciano. *Information: A Very Short Introduction.* Illustrated edition. OUP Oxford, 2010.

Green, Martin A., Ewan D. Dunlop, Jochen Hohl-Ebinger, Masahiro Yoshita, Nikos Kopidakis, Karsten Bothe, David Hinken, Michael Rauer, and Xiaojing Hao. "Solar Cell Efficiency Tables (Version 60)." Progress in Photovoltaics: Research and Applications 30, no. 7 (2022): 687–701. https://doi.org/10.1002/pip.3595.

Halliday, David, Robert Resnick, and Jearl Walker. *Fundamentals of Physics.* Extended 7th edition. Hoboken, NJ: Wiley, 2004.

Hubka, Vladimir. *Principles of Engineering Design.* Butterworth Scientific, 1982.

INCOSE. *INCOSE Systems Engineering Handbook: A Guide for System Life Cycle Processes and Activities.* 4th edition. Hoboken, New Jersey: Wiley, 2015.

Juvinall, Robert C., and Kurt M. Marshek. *Fundamentals of Machine Component Design.* 7th edition. Wiley, 2020.

Ogot, Madara, and Gul Okudan-Kremer. *Engineering Design: A Practical Guide.* Trafford Publishing, 2004.

Otto, Kevin, and Kristin Wood. *Product Design: Techniques in Reverse Engineering and New Product Development.* 1st edition. Upper Saddle River, NJ: Pearson, 2000.

Pahl, Gerhard, W. Beitz, J. Feldhusen, and K. H. Grote. *Engineering Design: A Systematic Approach.* Edited by Ken Wallace and Lucienne T. M. Blessing. 3rd edition. London: Springer, 2007.

Shannon, C. E. "A Mathematical Theory of Communication." *The Bell System Technical Journal* 27, no. 3 (July 1948): 379–423. https://doi.org/10.1002/j.1538-7305.1948.tb01338.x.

Shishko, Robert. *NASA Systems Engineering Handbook.* National Aeronautics and Space Administration, 1995.

Stoll, Henry W. *Product Design Methods and Practices.* CRC Press LLC, 1999.

Tu,J.F.,"Nuggets of Mechanical Engineering–Revisit of the Free- Body Diagram Analysis and Force Flow Concept," *Proceedings of the International Conference on Engineering Education* – ICEE 2007, Coimbra, Portugal September 3 – 7,( 2007).

Ullman, David. *The Mechanical Design Process*, 1st edition, McGraw-Hill, (1994). p206.

United States Department of Energy (DOE) "Furnaces and Boilers." Accessed August 1, 2023. https://www.energy.gov/energysaver/furnaces-and-boilers.

Young, Paul. *The Nature of Information.* First Edition. New York: Praeger, 1987.

# How It Works: Components and Subfunctions 3

## 3.1 Chapter Overview

- A technological system accomplishes its overall function by means of its constituent parts or components.
- The overall function of a technological system is divided into a network of subfunctions within the system.
- Components are the physical elements that embody the subfunction transformations. Components interact with other components.
- The interaction of two components requires the exchange of the same type of material, energy, or information.
- One approach to describing how a technological system works is to identify the major components, the subfunctions carried out by these components, and the principal component interactions that transform system inputs to outputs.
- Some components or combinations of components may have the function of controlling system behavior or operation.
- In some instances, the terms assembly or subassembly may be used to describe combinations of components that accomplish well-defined functions but do not appear as a completed system.
- Established components exist to provide some commonly used well-defined subfunctions. The components can be used to achieve desired subfunctions in any technological system.
- Component characteristics can be varied or adjusted to match system requirements.
- Operation of a technological system involves functions of both critical and less-critical contributions to overall system function.

© The Author(s), under exclusive license to Springer Nature Switzerland AG 2024 75
J. Krupczak, Jr., *Understanding Technological Systems*, Synthesis Lectures
on Engineering, Science, and Technology, https://doi.org/10.1007/978-3-031-45441-7_3

- Through an assemblage of components, a technological system can provide a function or utility that is not possible for any single component to accomplish. The combination produces a result that exceeds the usefulness of the individual parts.

## 3.2    Subfunctions and Components

System function is provided by components combined into systems. The overall function of the device, or technological system, is accomplished via subtasks or subfunctions. In a technological system, inputs crossing the system boundary are transformed into outputs. The overall functions of a technological system are accomplished through a network of subfunctions occurring within the system. Subfunctions contribute to the transformation of inputs to outputs. The subfunctions are the subproblems or subtasks which when combined, solve the main problem.

Figure 3.1 depicts several different technologies. One is a pharmaceutical manufacturing plant representing the type of facility that produces consumer medications. Another shows the inside of a mobile phone. The third is a kidney dialysis machine that is used for patients with kidney disease.

All of these systems transform available inputs into desired outputs. The pharmaceutical plant converts the ingredients into the final desired medication. The mobile phone transforms the input signal in the form of an electromagnetic wave to the desired audio

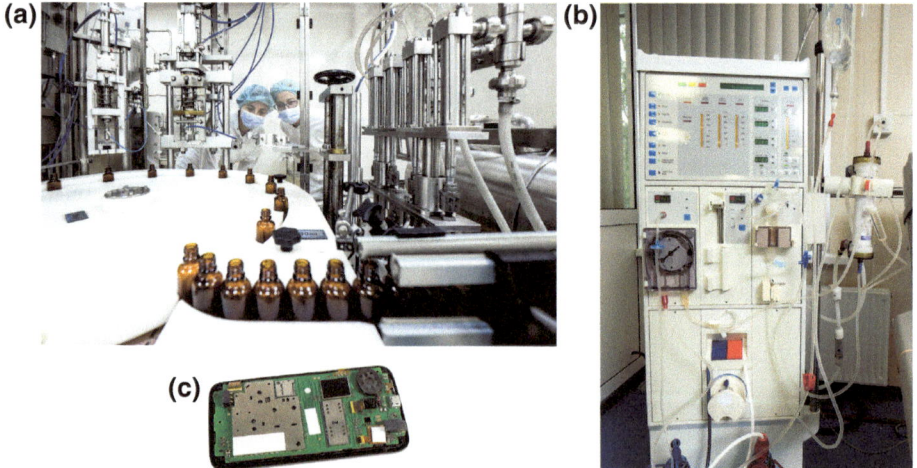

**Fig. 3.1**  Technological systems: **a** pharmaceutical manufacturing plant (iStock.com/extreme-pho tographer); **b** kidney dialysis machine (iStock.com/OlegMalyshev); **c** inside of a mobile phone (iSt ock.com/ekipaj)

and video output that can be understood by the user. The kidney dialysis machine receives an input of blood containing waste products and removes that waste normally removed by the kidneys and returns cleaned blood to the patient.

Images of a pharmaceutical plant, kidney dialysis machine and mobile phone show these all have the form of a combination of constituent pieces. The overall whole is created from an arrangement of component parts. The transformations of inputs to outputs accomplished in these systems are made possible by the combined action of the internal parts. This demonstrates a characteristic feature of technology, the net effect results from a network of interacting elements. The complex whole emerges from the combinations of the internal interactions of elements. These examples illustrate that technological systems are made from components which provide the subfunctions in support of accomplishing the overall function. Technological systems achieve transformations and accomplish their function using components internal to the system.

### 3.2.1  Labeled Directed Graphs

A system is a network of interacting elements to achieve a common purpose. The mathematics of graphs are relevant to systems composed of interacting elements. Mathematically, graphs are structures used to indicate the existence of relations between pairs of objects. A graph is made up of nodes or vertices which are connected by edges or links. The edges or links represent relations between the nodes. If the direction of links is considered significant, this is called a directed graph or a digraph. Nodes and links can be labeled leading to a labeled digraph. Examples of a graph of nodes and links along with a labeled digraph can be seen in Fig. 3.2.

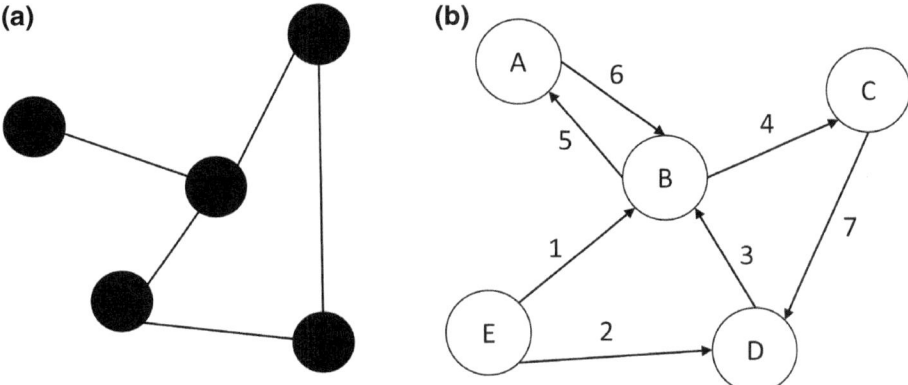

**Fig. 3.2**  A mathematical graph and a labeled directed graph (illustrations by author)

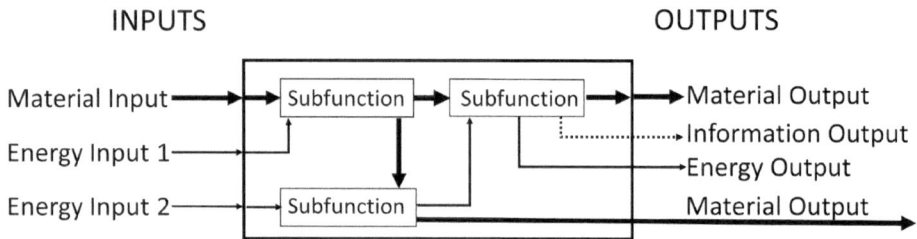

**Fig. 3.3** Technological system and subfunctions provided by components (illustration by author)

Graphs of nodes and links are frequently used to represent networks. Networks consist of nodes that link to other nodes. Examples of networks represented by this type of graph include social networks, structured data bases, and road networks.

The mathematical concept of a graph can be used to represent a technological system. Components providing subfunctions can be viewed as nodes. The interaction between components are the edges or links. Since the technological system transforms inputs into outputs, a direction exists in the system. The links have direction and labels.

Figure 3.3 shows a generalized form for a representation of a technological system using nodes and directed links. The nodes are components that carry out subfunctions in the system. The links are interaction between the components. These have direction and labels. Different arrow types are part of the labeling and can be used to distinguish interactions in the general form of material, energy, and information.

Subfunctions are responsible for transforming some subset of the inputs into a subset of the outputs. Intermediate inputs and outputs which are internal to the system may be produced. Subfunctions can be expressed as verb—noun pairs. The arrangement of interconnected subfunctions could be termed the function structure or function architecture.

### 3.2.2  Components

Technological systems are made from components which provide the subfunctions in support of accomplishing the overall function. Technological systems achieve transformations and accomplish their function using components internal to the system. Components solve subproblems or provide part of the system function. The subfunctions shown in Fig. 3.3 are carried out by specific components. Some physical component or collection of components carries out each subfunction. In some instances, the terms "assembly" or "subassembly" may be used to describe combinations of components that accomplish well-defined functions but do not appear as a completed system.

Flows of material, energy, and information form the input and outputs of the components. Components can have material, energy, and information flows that are internal to the technological system and do not cross the system boundary. Components are the physical elements that embody the subfunction transformations. Components interact with other components. The interaction of two components requires the exchange of the same type of material, energy, or information. Two components can interact if the output of one component can be received as an input to another component.

## 3.3    From How It Looks to How It Works

A description of how the system works can be developed from a form representation of how the system looks. Recognition that it is the components internal to the technological system that are responsible for transforming the inputs into the outputs, allows an analysis of "how it works" to be carried out starting from "how it looks."

"How it looks" refers to the form representation of a particular technological system. This is naturally the visual appearance of the system. The visual appearance does not necessarily convey useful information about how the system actually accomplishes its functions.

"How it works," refers to a description of how the system inputs are transformed into the outputs. Of specific concern are the components responsible for the subfunctions within the system, the interaction between these components, and the structure or architecture of the component arrangements.

Describing how something looks or the visual appearance of an object would generally emphasize the major features and distinctive characteristics. Similarly, in describing how the system works, a general approach will focus on the major components and functions most responsible for transforming the inputs to the outputs.

A block diagram is one way to describe how the system works similar to that shown in Fig. 3.3. The emphasis is on identifying the constituent elements of the system and their interactions. Form features related to physical appearance are not emphasized.

## 3.4    Example a Hair Dryer System Analysis

This section describes an example analysis of how a hair dryer works starting from how it looks. A hair dryer provides a specific example. Figure 3.4 shows the hair dryer and a view of the components internal to the hair dryer. The function of the hair dryer expressed in verb-noun form is to "dry hair". The function expressed more generally might be heat an air stream. The user of the hair dryer can apply this heated air stream to a variety of purposes, drying hair is just one option. The mode of operation considered assumes that the hair dryer is on and running under normal circumstances.

The hair dryer is an energy conversion system that receives inputs of electrical energy and air, converts the electrical energy to heat and transfers the heat to the air. The main

**(a)**                **(b)**

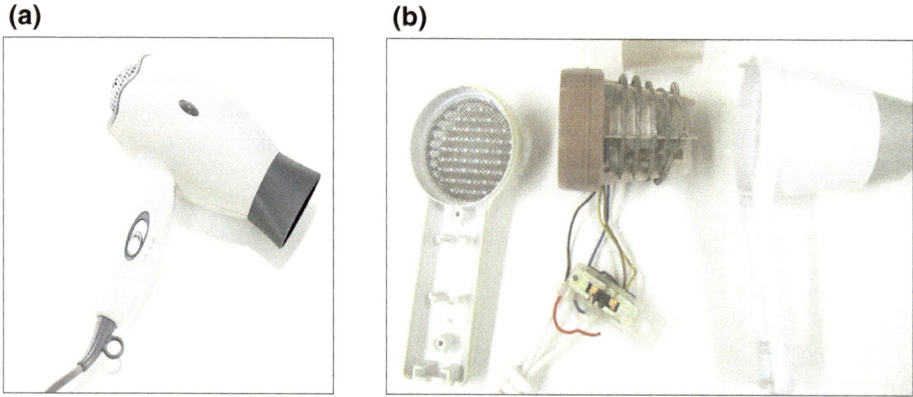

**Fig. 3.4** Hair dryer and internal components: **a** hair dryer (iStock.com/gerenme); **b** hair dryer components (photo by author)

transformations taking place in the hair dryer appear in Fig. 3.5. The inputs are air at room temperature and electrical energy provided by the electric current. The outputs are heated air and information in the form of noise or other indications that the hair dryer is operating.

Figure 3.5 depicts the end users' view of the hair dryer. The user expects to provide the dryer with electricity and in return the dryer will produce a flow of heated air. In the most basic consumer perspective, how the dryer accomplishes this transformation of electricity into hot air is not important. What it does matters, how it does it, does not matter.

The components of the hair dryer accomplish this transformation of the inputs into the outputs. The action of the components is how the hair dryer, or any technological system, transforms the inputs into the outputs.

*Hair Dryer Overall System Function*

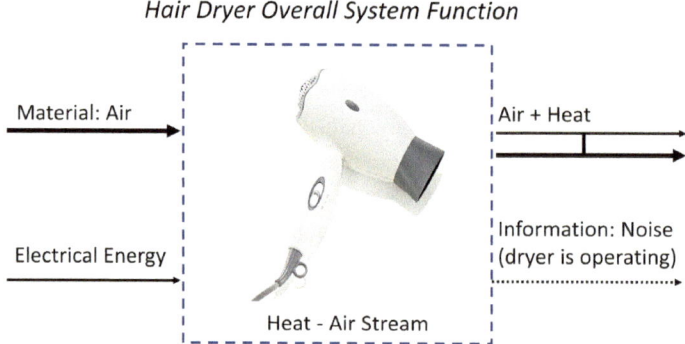

**Fig. 3.5** Hair dryer overall system function (iStock.com/gerenme)

Figure 3.6 shows a block diagram functional representation of a hair dryer. While the dryer is composed of dozens of individual pieces, it is possible to identify a group of components that are most important for accomplishing the overall function of the device. Energy is an input which enters the system as an electric current via the cord. The switch actuates the current and allows the current to reach the other components. The electric motor converts electrical energy into energy of motion which rotates the fan. The fan transfers energy of motion to the air creating flow of air over the heating coils. Electric current is converted into sensible heat by the heating coils. This heat energy is then transferred to the air. The nozzle channels the airflow, and the heated air is now an output of the system. Information is output in the form of noise from the fan which indicates that the dryer is operating as does the temperature of the output air stream.

*System Block Diagram Emphasizing Component Functions*
The representation of the hair dryer shown in the lower portion of Fig. 3.6 is a representation emphasizing the interacting functional elements of the hair dryer as a technological system. The diagram summarizes how the hair dryer works.

The block diagram identifies the major components and emphasizes the functional transformations accomplished by these components. The appearance or form of the component is deemphasized. The functional representation is an abstract depiction of the system describing the functional arrangement achieved by the system components. The diagram is not intended to show how the system looks but rather how the system works. The functional diagram conveys how the inputs are converted into the system outputs through interacting components performing specific intermediate steps.

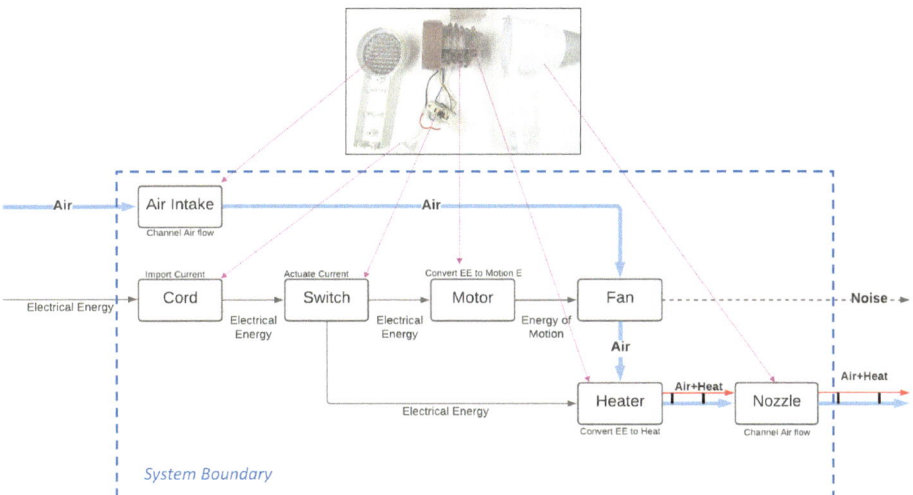

**Fig. 3.6** Block diagram functional representation of hair dryer components (photo by author)

The block diagram of the system is useful for identifying major components of the technological system and the interactions between these components during a specified mode of system operation. Components are represented as boxes. The inner workings of each component are not considered. In this way each component is treated like a system itself. Processes and effects within the component are not included at this point (The internal physical effects within components are discussed in Chap. 4).

## 3.5    Standard Components: Premade Functions

Components exist for commonly needed subfunctions. Established components exist to provide some well-defined subfunctions. The components can be used to fulfill desired subfunctions in any technological system. A system is not likely to be constructed entirely of standard components. However, recognition of established or standardized components facilitates understanding how a technological system works. The characteristics of standard components might be varied or adjusted to match system requirements (Components are treated in more detail in Chaps. 4 and 6).

Established components are listed below. These components are show in Fig. 3.7.

- Switch: The switch functions to actuate or allow the flow of electrical energy into the system.

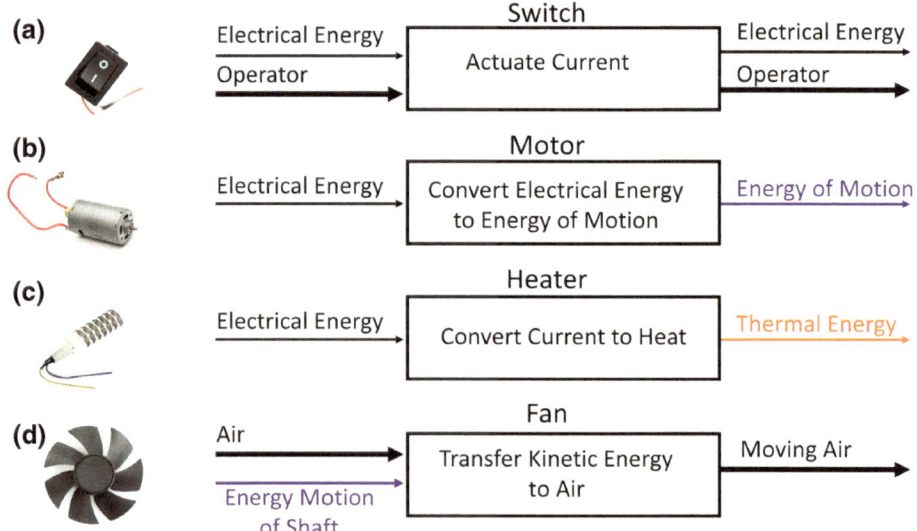

**Fig. 3.7** Standard components: **a** switch (iStock.com/Tamar Dundua); **b** electric motor (iStock.com/Antagain); **c** heater (iStock.com/Denis–Dryashkin); **d** fan (iStock.com/Sanny11)

- Electric Motor: The electric motor converts electrical energy into energy of motion.
- Fan: The fan transfers energy of motion from the fan blades to the air to create a flow of moving air.
- Heater: An electric heater converts electrical energy into thermal energy.

Classes of products utilize families of components. For example, most hair dryers of a similar type tend to utilize the same types of components. Technological groups or domains emerge as a set of solutions or design elements. Similar problems addressed by recurring functions lead to common forms or sets of component families and functional modules. Technological domains are considered in more detail in Chap. 8.

## 3.6    Primary and Secondary Functions and Critical Path

Some functions are more important than others and establish a "critical path" between the inputs and the outputs. In each mode of operation technological systems typically have a primary function that is the essential need of the user in that operating mode. In addition to this primary or principal function, technological systems may have components that carry out subsidiary or supplementary functions. These secondary functions contribute to the performance and utility of the system but are in some way subordinate or auxiliary to the main function in that operating mode.

In most instances a technological system can be identified as having an overall, main, or principal system function. This might be what is considered as the essential purpose of the device. If the system failed to perform this function the consumer or end user would notice and would consider the system to be non-operational or even worthless.

For example, it could be considered that the function of the automobile is to transport the driver and passengers. If the system is unable to carry out this function, the end user will certainly notice and consider the system to be non-functional. So, if a car does not start it cannot perform its main function. However, if a turn signal is not working the car can still carry out its main function. The turn signal performs a useful and helpful function, but it is possible to consider it to be in some way subsidiary to the main functions of the system.

Components responsible for the main function establish a "critical path" for flows of material, energy, and information through the device. If any of these vital components fails to carry out its function, then the overall operation of the system is compromised. It is most important to identify the components on the critical path when understanding the operation of the system.

Subsidiary functions contribute to operation but not the principal operation of the system. These auxiliary components often provide useful features.

The hair dryer shown in Fig. 3.8 serves as an example of subsidiary functions. Like most electrical devices, the hair dryer has an "ON" light which is illuminated when the

**Fig. 3.8** Hair dryer with an
ON light (iStock.com/ampols)

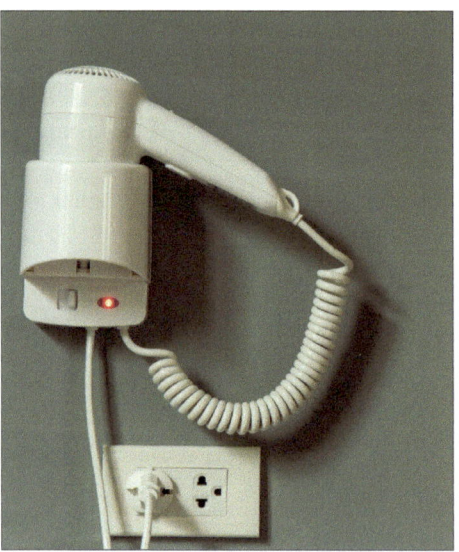

device is operating. The ON light can be considered a subsidiary function. It is not necessarily on the critical path of the main flow of air through the hair dryer. If the bulb
of the ON light stops working the hair dryer might still operate and perform its primary
function.

## 3.7    Component Secondary Inputs and Outputs

Components can have inputs and outputs that are disregarded in a system. In addition
to what might be considered the inputs and outputs of a component, secondary inputs
and outputs also occur. These secondary flows or effects are transformations of material,
energy, and information in a component but are of a magnitude such that they do not normally affect primary behavior at a level noticeable by the end user. In an initial analysis
of a technological system the secondary inputs and outputs are usually disregarded. However, these secondary flows may become of primary importance in some circumstances.
It is important to be aware of the existence of secondary effects even though they are
usually omitted for sake of clarity.

An electric motor is shown in Fig. 3.9. The function of the motor is to convert electrical
energy into kinetic energy. These are the primary inputs and outputs. However, the electric
motor also produces other outputs. Heat is produced along with noise and vibrations. In
addition, the motor requires some type of force to hold the motor fixed during operation.
The force is usually provided by some type of mounting structure.

These secondary inputs and outputs are usually considered to be of lesser importance
than the primary ones. They are not usually the reason the component is included in the

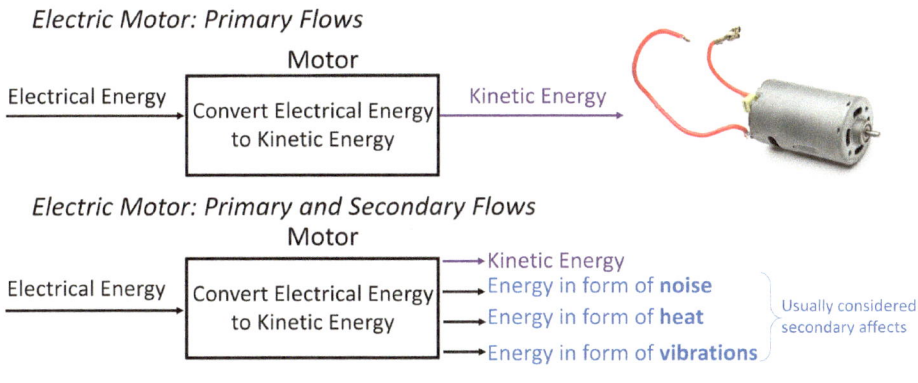

*Electric Motor: Primary Flows*

**Motor**

Electrical Energy → | Convert Electrical Energy to Kinetic Energy | → Kinetic Energy

*Electric Motor: Primary and Secondary Flows*

**Motor**

Electrical Energy → | Convert Electrical Energy to Kinetic Energy | →
- Kinetic Energy
- Energy in form of **noise**
- Energy in form of **heat**
- Energy in form of **vibrations**

Usually considered secondary affects

*Secondary functions can affect primary function in some circumstances*
- Motor too noisy
- Motor overheating
- Motor vibrations excessive

**Fig. 3.9**  Electric motor primary and secondary flows (iStock.com/Antagain)

system. However secondary inputs and outputs cannot be completely forgotten as they can affect primary function in some circumstances. For example, the motor might become too hot and stop functioning due to overheating. In some applications the sound emitted by the motor may be too noisy or the motor vibrations excessive for a particular application. If the mounting force is not sufficient the motor may loosen and cease to provide adequate torque as expected.

## 3.8    Feedback and Control

Not all system component interactions are directed from the input toward the output. Technological systems can have interactions that loop back from near the output to the input end of the system. In some cases, components or combinations of components may have the function of controlling system behavior or operation. These "feedback loops" can fulfill a control function in the system.

As an example, consider the basic hair dryer system with the addition of a component that prevents the exiting air flow from getting too hot and possibly injuring the user. A temperature-activated safety switch is included. Figure 3.10 depicts the system block diagram.

The thermal switch senses the temperature of the air exiting the heater. If the air exceeds the pre-determined maximum allowable temperature, the switch opens to stop electrical energy from reaching the heater.

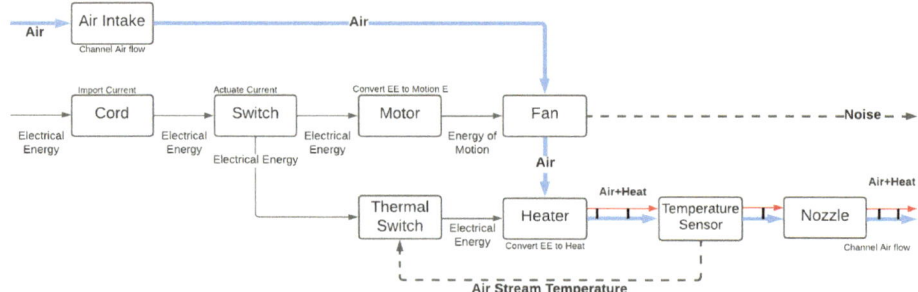

**Fig. 3.10**  Hair dryer with feedback control

## 3.9    The System is Greater Than the Sum of Its Parts

Combining components into a system enhances their utility. One perspective on technological systems is the creative combination of components produces a result that is more than the sum of the individual parts. The system components provide well-defined subfunctions but their usefulness is enhanced by the integration of components to achieve a function and utility that exceeds that provided by the components individually.

Consider the example of the hair dryer and the four main components of the switch, electric motor, fan, and heater as shown in Fig. 3.11. Each of these components can be used alone to provide a function. The switch actuates the flow of electric current. This is a useful function, but a switch alone is only a current actuator. The electric motor has an input of electrical energy and produces an output of a spinning shaft. This is an interesting phenomenon to observe but the utility of a motor by itself is limited. The heater can convert electrical energy to thermal energy. This thermal energy typically will be transferred to the air in the vicinity of the heater coils resulting in localized hot air. Hot air around coils of wire has minimal practical value. A fan produces flow of air. Various applications can be envisioned for a flow of air but untreated air at room temperature has limited uses.

When the components of the switch, motor, fan, and electric heater are creatively combined into a system it might be said that the net function of the system exceeds the function accomplished by the individual components operated independently.

## 3.10    Approach for Describing Technological Systems

A useful methodology for describing technological systems can be outlined. For understanding technology, it can be helpful to create basic and relatively uncluttered diagrams of major system components and their interactions. This section suggests an approach to use for transforming "how it looks" to "how it works."

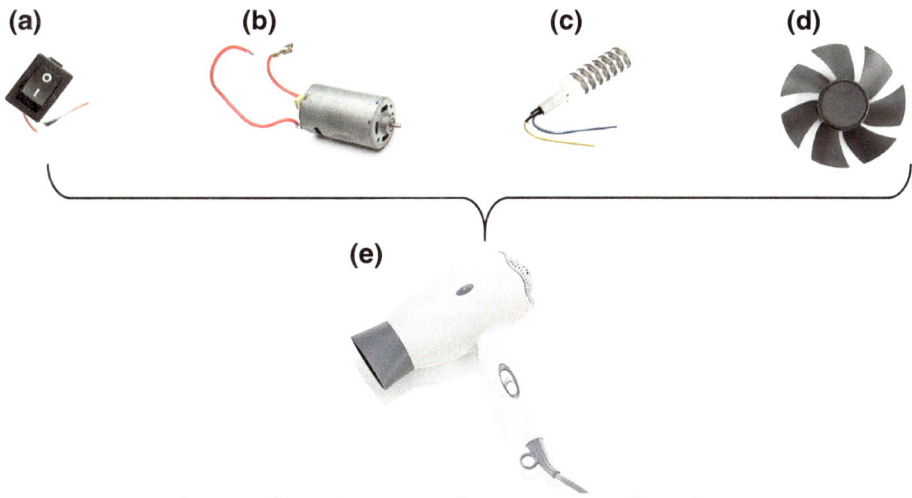

System function exceeds component functions.

**Fig. 3.11** System function exceeds component functions: **a** switch (iStock.com/Tamar Dundua); **b** electric motor (iStock.com/Antagain); **c** heater (iStock.com/Denis-Dryashkin); **d** fan (iStock.com/Sanny11)

The physical world presents itself to human senses in terms of form features. Human senses perceived form or the physical attributes of an object such as shape, color, and texture. "How it looks" is a shortened reference to the range of inputs to the human senses. The human senses can perceive some of the inputs and outputs and the physical attributes of a technological system but, for many types of technological systems, they are not well-suited for immediately perceiving how the inputs are transformed into the outputs or how the system works.

The mathematical concept of a graph of nodes and links representing a network, provides guidance about a procedure that might be used to describe the system. The task is to identify the nodes and their interconnections. For technological systems made by people it is relatively easy to identify the node and links as the network of components and their interactions that has been deliberately created by people. The nodes and interactions of actual physical hardware that has a well-defined and known purpose are much more readily recognized than most other types of systems. A challenge in technological systems which can contain numerous individual components is determining the level of detail at which to assign and identify components with nodes.

A graph of a network, like any mathematical model, is only an approximation of the physical world that can be useful for some purposes. Representing a technological system as a directed labeled graph approximates the behavior and structure of the real system in the same way that a drawing of the form of the system approximates the physical appearance. The drawing might include edges, shading, and selected detail to sufficiently convey how the system looks. Similarly, an individual describing an incident or telling a story relates what are considered to be the aspects most relevant to the point being made. Creating any type of representation of the physical world requires discerning the aspects most relevant to conveying a sense of the whole for the intended purposes.

Describing the system necessarily begins with the form representation of the technological system.

As an example, consider the kidney dialysis machine. The kidney dialysis machine, also called hemodialysis, is a filtering machine that removes wastes and excess fluids from the blood. It replaces the function of the kidney in the event of kidney failure. Figure 3.12 shows a typical kidney dialysis machine.

The objective is to describe the kidney dialysis machine as a technological system in terms of the constituent components and their interactions converting the available inputs into the desired outputs.

### 1. *System Boundary*

Define the system boundary. The system boundary separates the system from the environment. The boundary determines what is within the system and what is external to the system. The system boundary is significant because the inputs and outputs to the system are identified at the system boundary. The system boundary is like an imaginary box that encloses the system.

**Fig. 3.12** Kidney hemodialysis machine [image cropped] (iStock.com/sompon g_tom)

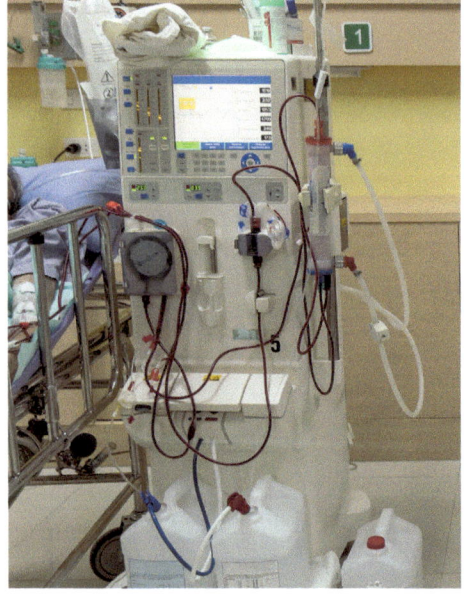

Example:

The machine is considered to be the system. In this case the patient is outside the system. The kidney dialysis machine interacts with the patient but the patient is external to the system.

### 2. *System Function*

A next step is to describe the primary function of the system in the mode of operation under consideration. This should be the highest-level overall function. This main function can be stated in verb-noun form. If a system is used for multiple independent high-level functions, these should be considered separately as separate functions. It is helpful to state the overall function in terms that have the fewest assumptions.

Example:

The primary function of the kidney dialysis machine is to clean patients' blood. More specifically, the function of the system is to remove excess water, solutes, and toxins from the blood.
    Function: Remove—water, solutes, toxins from blood.

### 3. *Mode of Operation*

Most technological systems have multiple different phases or modes of operation. In attempting to describe how a system works the mode of operation under consideration must be specified. For example, a washing machine has a fill cycle and a wash cycle. Different inputs and outputs may occur in different operating modes. The interaction between components can be different in different aspects of system operation. A time period over which the system operates may be specified if relevant.

Example:

The kidney dialysis machine is in operation cleaning the blood of a patient.

### 4. *Inputs and Outputs*

System inputs and outputs are identified at the system boundary. The inputs move into the system from some source outside of the system. The outputs leave the technological system. The primary function of the system in question is to transform these available inputs into the desired outputs.

Example:

A primary input to the kidney machine is blood containing waste products. A primary output is cleaned blood. During operation, solutes and toxins from the blood are transfer to a dialysate fluid. Fresh unused dialysate fluid is an input. Used dialysate fluid containing the wastes removed from the blood is an output that leave the system. Electrical energy is another major input to this technological system. Various displays provide information regarding operational status of the system.

5. *Classification and Overall View*

For technological systems it can be helpful to classify the inputs and outputs in terms of material, energy (or related concepts of power and force) or information. For many technological systems, energy in some form must be an input. Information may enter and leave the system.

Example:

For the kidney dialysis machine, the blood and dialysate fluid are materials. The system also uses electrical energy. Figure 3.13 is a summary of the overall operation of the system.

6. *Critical Path Components and Level of Detail for Description*

The components internal to the system carry out the subfunctions that convert the inputs to the outputs in a particular operating mode. In describing a technological system, a key is identification of the critical paths within the system that connect inputs to outputs. This step focuses on the components and their interactions.

Systems have multiple levels or hierarchies. Any given description must settle on the level to be represented. Determining the appropriate level of detail is an important issue in technological system descriptions. This will determine the approximate number of nodes in the form of components to be included. From a practical standpoint clutter diminishes the interpretability of diagrams. Starting from a form representation of how the system looks, most technological systems can be represented initially by about 10 (5–15) main components. If desired, greater levels of detail can be pursued in a layered manner identifying subcomponents but the goal is clarity in conveying how the system works.

It is essential to show enough detail to be informative and non-trivial, however excessive amounts of detail appear cluttered and obscure the meaning. Technological system

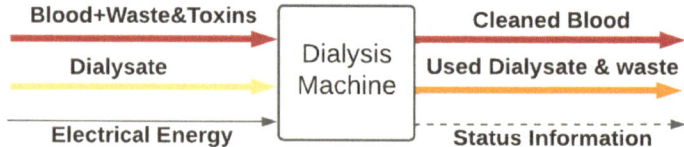

**Fig. 3.13**  Dialysis machine overview

representation can be pursued to more and more levels of detail just like a drawing of an object or a story describing an incident can be more and more detailed. A guideline of 5–15 components is intended to be an appropriate number for describing the components and subfunctions that form the critical path in achieving the overall system function. Figure 3.14 identifies the major components of the kidney dialysis machine.

The dialysis filter is the component in which the actual removal of waste and toxins from the blood takes place. A dialysis filter is shown in Fig. 3.15. The patient's blood is on one side of the filter and the dialysate fluid is on the other. Waste and toxins from the blood are transferred through the filter to the dialysate fluid.

The patient's blood is pumped through the system by the blood pump. A pump is also used to push the dialysate fluid through the filter. Monitoring units are used to measure blood pressure at critical locations in the system. The anticoagulant heparin is added to the blood through a separate pump. Heparin reduces the possibility of the patient's blood clotting within the system.

Several other major features inform a description of this system. The patient's blood is removed from an artery and returned to a vein. The heparin is added before the dialyzer. To inform the operator of critical system conditions, blood pressure is measured both when leaving and returning to the patient and at the input to the dialyzer. Bubbles and air in the blood are removed by a filter or trap immediately before being returned to the patient.

7. *Assemble the System*

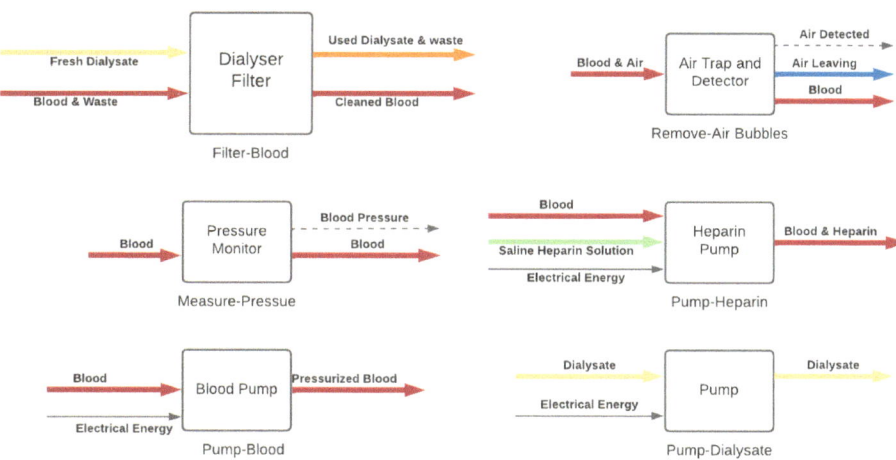

**Fig. 3.14**  Dialysis machine major components

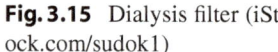

**Fig. 3.15** Dialysis filter (iSt
ock.com/sudok1)

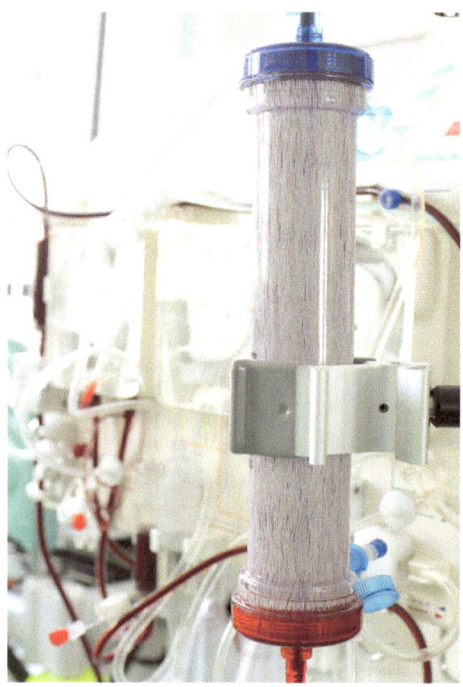

The component information is used to assemble a representation of the system in the form
of a node and link diagram. The components are the nodes. The links between nodes are
the inputs and outputs of each component. A goal is to connect the overall system inputs
to the overall outputs through a network of components carrying out subfunctions that
contribute to the transformation of the inputs to the outputs.

A representation of the kidney dialysis system is shown in Fig. 3.16. Blood containing
waste products is removed from a patient's artery. Blood pressure is measured and a
pump provides the pressure needed to push the blood through the system. A separate
pump adds anticoagulant heparin before the blood enters the dialysis filter. In the filter
waste products in the blood are transferred to the dialysis fluid. The used dialysis fluid
with waste products exits the system. Bubbles or air that may have become entrapped in
the blood are removed before the cleaned blood is returned to the patient.

The creative combination of components produces a system that achieves greater function than the individual parts can accomplish separately. Within the system components
carry out subfunctions and interact to achieve transformation of the available inputs into
the desired outputs. Through an assemblage of components, a technological system is
able to provide a function or utility that is not possible for any single component to
accomplish.

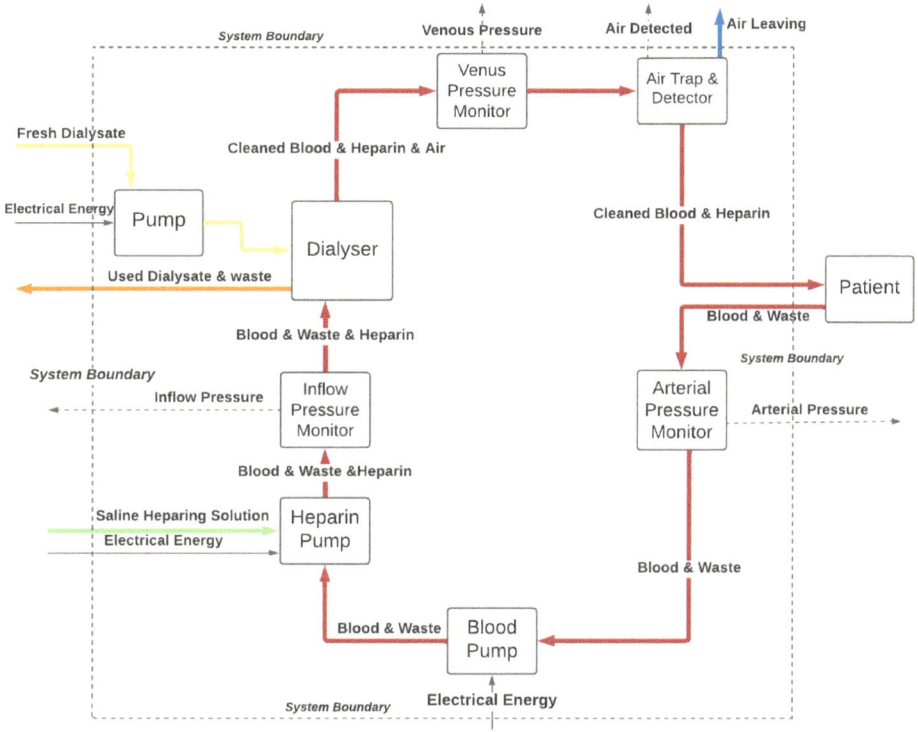

**Fig. 3.16** Diagram of kidney dialysis system

## Bibliography

Alexander, Christopher. *Notes on the Synthesis of Form.* Harvard University Press, 1964.

Azar, Ahmad Taher, and Bernard Canaud. "Hemodialysis System." In *Modelling and Control of Dialysis Systems: Volume 1: Modeling Techniques of Hemodialysis Systems*, edited by Ahmad Taher Azar, 99–166. Studies in Computational Intelligence. Berlin, Heidelberg: Springer, 2013. https://doi.org/10.1007/978-3-642-27458-9_3.

Buckley, Fred, and Marty Lewinter. *Introductory Graph Theory with Applications.* 1st edition. Waveland Press, Inc., 2013.

Dym, Clive L., and Patrick Little. *Engineering Design: A Project-Based Introduction.* Wiley, 2004.

Hatley, Derek J., and Imtiaz A. Pirbhai. *Strategies for Real-Time System Specification.* New York, NY: Dorset House, 1988.

Hubka, Vladimir. *Principles of Engineering Design.* Butterworth Scientific, 1982.

Otto, Kevin, and Kristin Wood. *Product Design: Techniques in Reverse Engineering and New Product Development.* 1st edition. Upper Saddle River, NJ: Pearson, 2000.

Pahl, Gerhard, W. Beitz, J. Feldhusen, and K. H. Grote. *Engineering Design: A Systematic Approach.*
    Edited by Ken Wallace and Lucienne T. M. Blessing. 3rd edition. London: Springer, 2007.
Ullman, David. *The Mechanical Design Process*, 1st edition, McGraw-Hill, (1994). p206.

# Phenomena and Models

**4**

## 4.1 Chapter Overview

- Components are created by people to provide a specific function in a technological system; however, all technology is actually part of the natural world thus the behavior of components is determined by natural phenomena.
- Components utilize natural phenomena in carrying out functions within technological systems. Although technology is made by humans, its behavior is governed by the same laws of nature that apply to the natural world.
- Science uncovers and clarifies the behavior of the natural world. The engineering sciences are those with particular relevance to the behavior of technological components and systems.
- The action or behavior of components is often described mathematically. Mathematical models or formulas describing component behavior can be based on the physical principles responsible for the characteristics of that component, empirical characterizations, or combinations of theoretical and experimental results.
- Mathematical models of component and system behavior typically have limits of applicability.
- The predictive capabilities of mathematical models facilitate component design, selection, interconnection, and the achievement of overall system performance.

© The Author(s), under exclusive license to Springer Nature Switzerland AG 2024    95
J. Krupczak, Jr., *Understanding Technological Systems*, Synthesis Lectures
on Engineering, Science, and Technology, https://doi.org/10.1007/978-3-031-45441-7_4

## 4.2    Components Utilize Physical Phenomena

In the utilization of physical effects in the engineering of components, the role of science and mathematics is evident in technology. Science investigates and analyzes the physical phenomena of the natural world. This understanding can be applied in creating and modifying components.

The action or behavior of components is often described mathematically based on the physical principles most responsible for the characteristics of that component. The mathematical models or formulas facilitate the development of technological systems by allowing predictions to be made of component behavior under expected conditions.

Components accomplish functions through use of physical phenomena. Technological components make use of natural phenomena. What is meant by the term "natural phenomena" are effects or behaviors that take place due to physical processes. Although technology is made by humans, the behavior of technological components is governed by the same laws of nature that apply to the natural or non-human made world. Components accomplish functions through use of physical phenomena. Components utilize and apply physical effects to achieve a specific purpose. Essentially components are "captive phenomena." Creation of technology is the manipulation of various natural phenomena to provide useful functions. The component can be viewed as the embodiment of an effect.

Utilizing effects does not require detailed understanding of the underlying physical behaviors used. The terms "effect" or "phenomenon" are used here rather than "principle" because principle implies some human understanding and articulation of the natural processes involved. It is not essential that the mechanism and principles underlying the phenomena utilized in a component be fully articulated and understood. In fact, in earlier technological eras, effects were utilized to accomplish desired functions without an understanding of the fundamental principles at work. For example, the use of water power preceded the mathematical development of fluid mechanics.

Presently, science investigates and analyzes the physical phenomena of the natural world. This understanding can be applied in creating and modifying components.

Essential component action can often be described succinctly through a few basic principles. Frequently, only one or two key phenomena or effects underly how the component carries out a transformation of inputs to outputs. Often there is a major physical effect responsible for providing the component function. In many instances components utilize only a limited number of fundamental processes. Essential component action can be described succinctly through a few basic principles.

Component form and phenomena utilized are closely related. Components utilize behaviors derived from underlying physical principles of the form. Components provide a specific function in technological systems by employing physical phenomenon and natural effects. The form of the component is selected, determined, and designed so that the natural phenomena associated with a specific arrangement of form provides a desired function. Specific form features are selected based on the intended phenomena or effects to be utilized.

Components vary in complexity. The natural phenomenon associated with a particular component does not have to be elaborate or complex for that component to be useful for solving an important problem. In other cases, the natural phenomenon employed in a technological component can be very complex. Components can also involve multiple effects simultaneously to achieve a desired function.

## 4.3   Examples: Components Applying Physical Effects

### 4.3.1   Component: The Electrical Switch

An electrical wall switch illustrates the relationship between underlying physical phenomena and component function and form. The wall switch is a common and relatively simple device. Figure 4.1 illustrates a common wall switch.

The switch provides a critical function in the operation of electrical devices, namely, to control if the device is operating by allowing electric current into the device. The switch actuates current.

Electrical wall switches use the phenomenon of electrical conductivity and the difference between good and poor conductors. They also employ the phenomenon of elasticity of metals.

Electrical conductivity is the property of materials allowing electric current in the material. Materials such as metals are good conductors and offer little resistance to electric

**Fig. 4.1** Common wall switch: **a** wall outlet (iStock.com/viafilms); **b** light switch component (iStock.com/joebelanger)

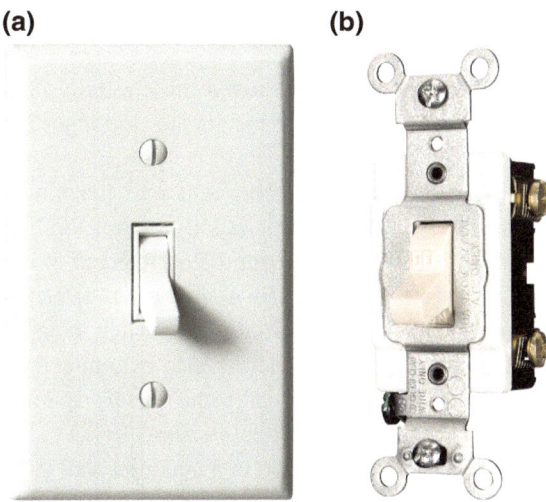

**(a)**                    **(b)**

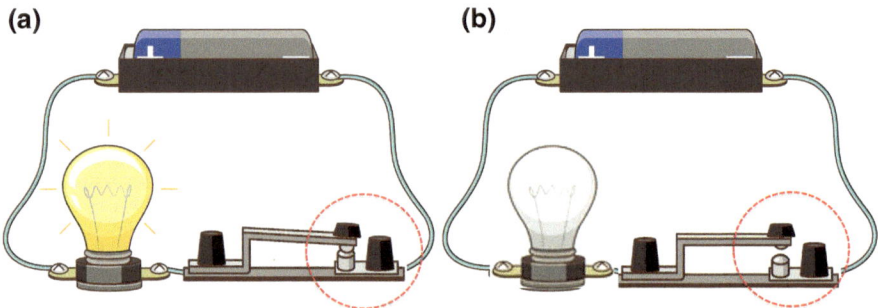

**Fig. 4.2** Basic electrical switch representation: **a** metals are in contact, electrical current is flowing, light is on (iStock.com/GraphicsRF.com); **b** metals are not in contact, no electrical current, light is off (iStock.com/GraphicsRF.com)

current. Electric current occurs easily in conductive metals. Other materials such as air are poor conductors. Electric current does not readily exist in poor conductors.

Elasticity is the springy property of many metals. If the metal is bent or deflected a small amount it tends to spring back to its original position.

Consider first a simple version of an electrical switch such as shown in Fig. 4.2. The electrical switch consists of two pieces of metal. When the switch is OFF the metals are not in contact and only air fills the space between the metals. Electric current can occur in the metals but not in the air. No complete electrically conductive path exists between the battery and the light, no current flows, the light is OFF.

When a force is applied and metals are brought into touching contact, the non-conductive air gap no longer exists. Both metals readily permit the flow of electric current, and the metals are in direct contact. Uninterrupted current can exist in the circuit. The light is ON.

When the force is removed the elasticity of the metal causes it to spring back to its original orientation. The metals are no longer in contact. The current stops and the light is OFF.

When the metals are brought into touching contact, the non-conductive air gap no longer exists. Both metals readily permit the flow of electric current, and the metals are in direct contact. Uninterrupted current can exist in the circuit. The light is ON.

The electrical switch is a simple but effective manipulation of form to carry out the function of actuating current. The switch is an application of the phenomenon of electrical conductivity and the difference in conductivity between metals and air. The elasticity or springiness property of some metals is used to easily and repeatedly connect and disconnect the metal switch elements.

The actual construction of a common wall switch differs from the simplified conceptual switch shown in Fig. 4.2, however, the actual implementation is not that different and the underlying principles used are the same.

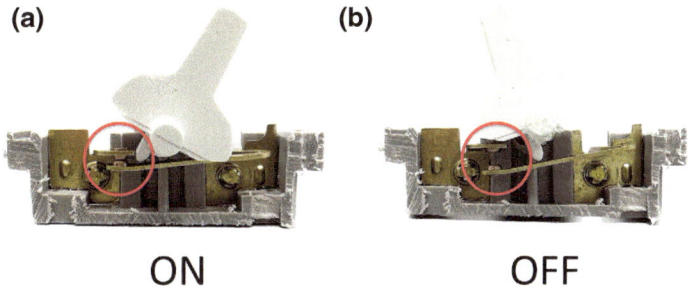

**Fig. 4.3** Internal components of an electric switch: **a** metals are in contact (photo by author); **b** metals are not in contact (photo by author)

Figure 4.3 shows internal views of a standard wall switch. This switch is based on the same principle and similar in form as the simple switch. When the switch is ON a thin piece of brass metal is in contact with a metal tab at the other end of the switch. There is metal-to-metal contact current can exist.

When the switch is turned OFF a plastic lever pushes the brass piece away from the metal tab. The elasticity of the metal allows it to spring back. Non-conductive air fills the gap. The switch is OFF.

The function of the switch is to actuate electric current. The phenomena utilized is the conductive property of metals in contrast to that of air. The form or physical characteristics of the switch are derived and linked to the utilization of the conductive property of metals. This type of switch must be made of a high-conductivity metal. The flexing behavior of the metal element between ON and OFF requires a metal with appropriate elastic behavior.

The basic form requirements of the switch are use of a metal that is conductive and elastic. These are two overriding essential features. A practical switch will have other requirements that influence the form of parts of this component. For example, the metal must also have appropriate size to allow the intended current to exist without substantial heat generation. The electrical properties of the surfaces in contact should not degrade due to chemical reaction with air or moisture. The property enabling flexing must remain consistent for thousands of on and off cycles over the life of the switch. Material and manufacturing costs for the component must be within acceptable ranges.

It should be noted that in the case of the switch the use of the known phenomenon of the conductivity of metals in a switch does not require any explanation of why this effect occurs in metals. The invention of this type of switch preceded the development of a theory for the conductivity of metals.

The original design of this type of switch does require any underlying theory of why metals tend to be good conductors. What is essential is a recognition of the effect or phenomenon of conductivity of metals compared to that of air. While a theory of why metals are conductive might be useful in designing this type of switch, it was not essential. The invention of the metal-to-metal contact switch preceded any consensus theory explaining the conductivity of metals.

Summarizing, the common electrical switch illustrates how physical effects are employed to achieve specific functions in technological components. If two electrical conductors are brought into physical contact, electric current can flow through the materials. When there is an air gap between the materials, no current flows. All contact-type switches employ this underlying phenomenon of electrical conductivity.

## 4.3.2 Component: Fractionating Column

The fractionating column is not a familiar everyday object but also illustrates the role of natural phenomena in components. A fractionating column is used to separate a mixture of liquids into its constituent parts. The production of many liquids essential to daily life rely on their operation. Fractionating columns are used extensively in industrial scale processing developing chemicals needed in solvents, detergents, adhesives, plastics, resins, and lubricants. Fractionating columns are also a key component in petroleum processing, natural gas processing, brewing, and the extraction of nitrogen and oxygen from air.

Fractionating columns are based on the difference in boiling temperatures of liquids. Different liquids have different boiling temperatures under the same pressure conditions. For example, under normal conditions water boils at 100 °C (212 °F) while alcohol boils at 78 °C. When a liquid boils it undergoes a change from being in a liquid state to a gaseous vapor state.

When a liquid mixture is heated the constituent liquid with the lower boiling temperature will tend to vaporize more readily. The vapor leaving the mixture has a greater fraction of lower boiling point component than the mixture itself. If this process is repeated for several stages, then one end result will be an output that is mostly comprised of the lower boiling point liquid. The vapor can be collected and cooled back into a liquid. Similarly, another output will be the high boiling point liquid from which the lower boiling point component has been driven off. In this way the mixed liquids can be separated due to their boiling point difference.

Figure 4.4 is a representation of the fractionating column; a multi-component liquid mixture is an input. This is represented by the Liquid A + B + C + D + E label to indicate that the liquid actually consists of multiple separate liquids mixed together. Energy in the form of heat is also input as this is required to heat the mixture to boiling or conditions to promote vaporization. The outputs are the separated flows of liquids A through liquid E. The individual component liquids separate out in successive order based on vaporization temperature. The heat energy added to the mixture to promote boiling and vaporization must be removed as the component liquids condense.

Figure 4.5 shows an image of an industrial-scale fractionating column. The fractionating column is an actual column. The mixture of fluids enters near the bottom where it

**Fig. 4.4** Fractionation
column separating liquids due
to different boiling
temperatures
(iStock.com/Sergey Merkulov)

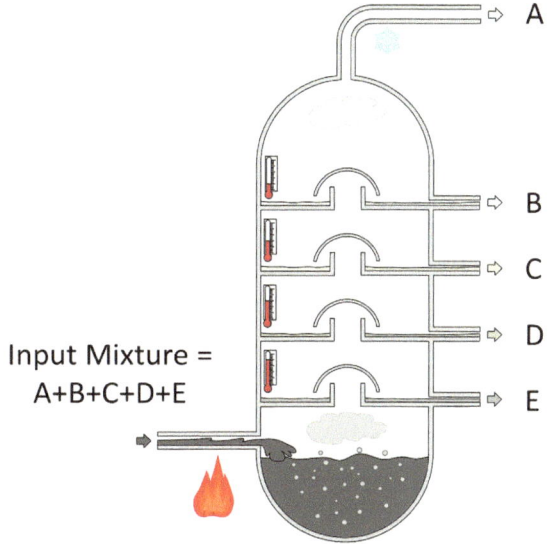

Input Mixture =
A+B+C+D+E

is heated. The heated vapor rises up the column. The higher up the column, the lower
the temperature and the greater the concentration of the lower boiling point liquid. The
vapor of mostly lower boiling point liquid exits the top of the column. Heat energy is
then removed from this vapor, cooling it to the liquid phase. The physical form of the
fractionating column is determined by the configuration needed to implement the effect
utilized.

An important application of the fractionating column is the production of petrochemicals and hydrocarbon fuels from crude oil. Crude oil as it is obtained from the earth
by oil drilling is a mixture of many different hydrocarbon liquids. This includes alkenes,
naphthenes and aromatics present in crude oil, which are used in processes to produce
essential chemicals such as solvents and plastics. Crude oil also includes different hydrocarbon liquids such as kerosene, gasoline, and diesel fuel. These liquid constituents of
crude oil have different boiling points. In refineries fractionating columns are used to
separate the constituents from the crude oil mixture.

Using the phenomenon of different boiling temperatures to separate liquids does not
necessarily depend on any understanding of why the constituent liquids have different
point temperatures or why the vapor leaving the heated mixture has a higher concentration
of the lower boiling point liquid. The phenomenon has been used by people for centuries
to distill alcohol. The distilling process for whiskey and other high-proof alcohol utilizes
a version of the fractionating column.

Figure 4.6 shows a fractionating column used in the separation of alcohol and water
in the distilling process.

The ancient application of the concept of a fractionating column in distilling alcohol
did not require an explanation of why the effect occurred, only that it existed and how it

**Fig. 4.5** Large industrial-scale
fractionating columns (iStock.
com/Prajakfoto)

could be used. Fractionating columns for early alcohol distillation were developed through observation and trial and error methods. Currently modern chemical engineering uses mathematics-based methods. The design of a fractionating column for a particular mixture of liquids is informed by an underlying theory of the thermodynamics of the equilibrium between phases.

### 4.3.3   Component: Electric Generator

One common type of electric generator employs the phenomenon of electromagnetic induction.

This type of generator converts energy of motion into electrical energy. This includes large utility-scale generators used in fossil fuel utilities such as coal, natural gas, diesel, and nuclear-fueled electric power, hydropower, as well as wind turbines. Electromagnetic induction is also used by smaller-scale gasoline and propane powered electric generators.

**Fig. 4.6** Copper fractionating column for alcohol distillation (iStock.com/Baloncici)

This type of generator utilizes the phenomenon of electromagnetic induction. In this effect when the amount of magnetic field in a conductor changes, an electric current is induced in the conductor. Typically, the conductor is a coil of wire. If a magnet near the coil moves, the amount of the magnetic field going through coil changes. This induces an electric current in the coil. Figure 4.7 shows this effect. The moving magnet induces a current in the coil.

Since the effect is derived from the changing magnetic field in the conductor, a current is induced whenever there is relative motion between the magnet and coil. A current is induced if the magnet is stationary, and the nearby coil is moving. A current is also induced if the magnet is moving and the coil is stationary. Rotating the coil is a convenient way to sustain motion. This is illustrated in Fig. 4.8. When a coil is continuously rotated in the magnetic field between the North and South poles of a magnet, an electric current is induced.

An expanded view of a typical practical generator configuration shown in Fig. 4.9. The shaft is rotated by various means such as a steam turbine, engine, or wind power. The coil sits between the poles of a magnet. As the shaft rotates the amount of the magnetic field

**Fig. 4.7** Electromagnetic
induction (Fouad A.
Saad/Shutterstock.com)

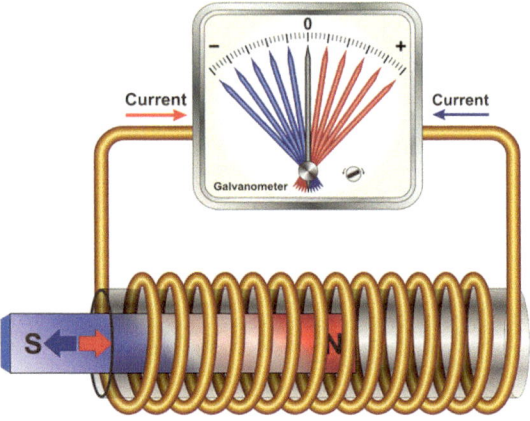

**Fig. 4.8** Simple electric
generator using
electromagnetic induction (ter
setki/Shutterstock.com)

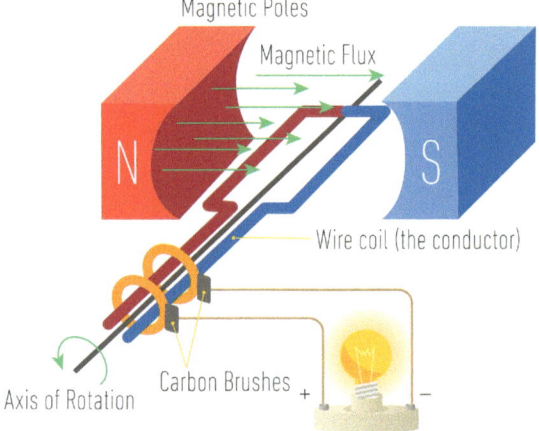

going through the coils changes inducing an electric current in the coils. The induced
current from the coils provides electric power to systems that require electrical energy.

The energy of motion is the input and electrical energy represented by the current is
output. The input is energy to spin a shaft. The output is electrical energy.

In the case of the electric generator, the discovery of the phenomenon of electromagnetic induction occurred first. The generator followed from the discovery of the effect.
This is an example of the understanding of a physical principle leading to the development
of a practical component based on the effect.

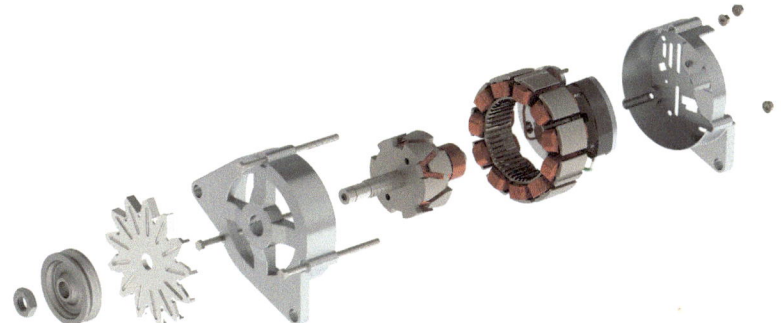

**Fig. 4.9**  Expanded view of a typical practical generator (BVNIMA/Shutterstock.com)

## 4.4     Mathematical Models of Technological Components

### 4.4.1   Mathematical Models from a System Component Perspective

A mathematical model is a representation of the physical behavior of a component described using a mathematical analogy. The characteristics and parameters of the physical situation are represented by terms or variables in a mathematical equation. The equation describes the relationship between the physical variables. The usefulness of mathematical models is they can be used to predict the behavior of a physical system. Mathematical models can be used in determining both form and function properties of components.

This section is intended to provide one perspective on the role of mathematics in understanding technological systems. The goal is to briefly describe in broad outlines the relationship of mathematics to the view of technology as systems of components. A characteristic of all modern engineering fields is the use of mathematics in the development of technological systems. Consequently, a comprehensive cataloging of the many applications of mathematics in engineering is beyond the scope of this work, as it is already well-treated in the extensive current engineering science literature.

### 4.4.2   Models Addressing Functional Performance

Some models relate component functional performance to component inputs and form parameters. The model can be used to determine a quantitative value for a component output in terms of input values and characteristics of form of the component. The model allows variation of specific input and form conditions to realize a desired quantitative output.

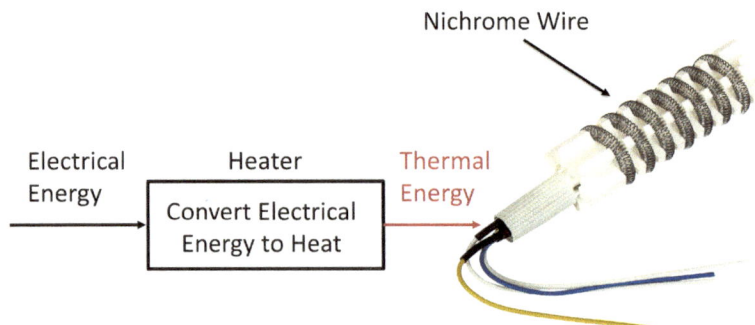

**Fig. 4.10**  Electric heater (iStock.com/Denis_Dryashkin)

*Electric Heater*

An electric heater converts electrical energy into heat. One type of electric heater is based on the heat produced by electric current in a wire. Figure 4.10 shows a typical electric heater component.

Joule's law relates power dissipated as heat to the electrical current and resistance in a conductor. A mathematical expression of Joule' law describes the heat produced $P$ by an electrical current $I$ in a conductor of electrical resistance $R$.

$$P = I^2 R$$

A mathematical model for the electrical resistance $R$ of a wire is:

$$R = \rho \frac{L}{A}$$

The formula expresses the property of electrical resistance in terms of other form properties. Here $L$ is the length of the wire and $A$ is the area of the wire cross section. The parameter $\rho$ is resistivity which is a material property of the metal used in the wire.

These two mathematical models can be combined resulting in the formula:

$$P = I^2 \rho \frac{L}{A}$$

This provides an expression for the desired component output of heat in terms of the input current and component form properties of length, cross sectional area, and material resistivity. If a particular amount of output heat is desired, the input and form parameters can be adjusted to obtain the desired output.

*Heater Design*

As an example, suppose it is desired to create a heater that will produce 100 Watts. The heater will use nichrome wire with a diameter of 1.0 mm and resistivity $\rho$ equal to $1.5 \times$

$10^{-6}$ $\Omega$m. The available current $I$ is 10 A. What length L of wire is needed to achieve 100 W output?

The equation for power $P$ can be rearranged to solve for length $L$.

$$L = \frac{PA}{I^2\rho}$$

For wire with a circular cross section the area A can be found from the equation for area of a circle.

$$A = \frac{\pi}{4}D^2$$

Using the known diameter D = 1 mm = 0.001 m, the area A can be found.

$$A = \frac{\pi}{4}D^2 = \frac{\pi}{4}(0.001)^2 = 7.85 \times 10^{-7}\,\text{m}^2$$

The length L can now be determined.

$$L = \frac{PA}{I^2\rho} = \frac{100\,\text{W} \times \left(7.85 \times 10^{-7}\,\text{m}^2\right)}{(10A)^2 \times \left(1.5 \times 10^{-6}\,\Omega\text{m}\right)} = 0.52\,\text{m} = 20.5\,\text{inch}$$

The analysis determined that to achieve the desired transformation of input electrical energy to output of heat, the length available wire needed is 0.52 m (20.5 inches).

An advantage of a mathematical model is the ability to use the mathematical equation to interpret the influence of component parameters on behavior. The mathematical model for heater power was found to be:

$$P = I^2\rho\frac{L}{A}$$

Consider the effect of changing one of the parameters while the others remain constant. From this it can be seen that the heat produced depends on the square of the input current $I$. So increasing $I$ has a significant impact on heat produced. A higher resistivity $\rho$ of the heater material results in more heat produced. A longer heater wire $L$ also increases heat output. Interestingly a larger diameter of wire with a larger cross sectional area $A$ results in less heat produced. This can be understood by noticing that a larger area A results in lower heater resistance and therefore a decrease in heat produced according to Joule's Law.

### 4.4.3  Models Addressing Form Characteristics

Mathematical models can be used to determine form properties or conditions under given component functional performance conditions to establish if requirements pertaining to form are being met. In this situation the functional parameters are known, and it is necessary to determine if a particular form requirement is achieved. Form requirements are specifications that are not associated with the component function but are imposed on the form properties of the system. For example, weight, size and cost are form-related requirements.

*Example: Beam bending*
A beam is a component used to support and transfer loads in a structure. An example of a steel beam structure is shown in Fig. 4.11. Beams transfer loads in the structure to the foundation.

   As a simplified example consider just a single beam resting on supports at the ends. The beam transfers load F to the supports at points A and B. The load is applied at the mid-point of the beam between the supports. The beam can be considered as having an input load F and output loads at A and B. Experience indicates that the beam will bend due to the force F. Figure 4.12 illustrates the case of the simplified beam.

**Fig. 4.11**  A steel beam structure (iStock.com/zhengzaishuru)

**Fig. 4.12**  A simply supported beam with a force F applied at the mid-point (iStock.com/Kolonko)

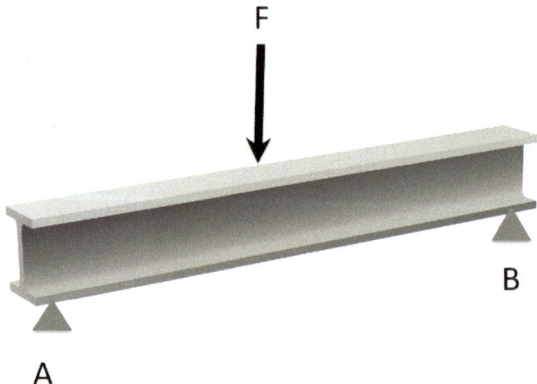

In this case a requirement is imposed on the amount of bending or deflection from unstressed position. The deflection must be within some tolerable amount. This is a requirement on the form of the component. During use, geometric dimensions of the component must have a specific property.

The deflection is a condition on the form of this component. Typically, a design requirement might be to transfer a desired load with deflection lower than a desired limit. Deflection is not a function; it is a condition on the form of this component during operation.

A mathematical model exists for the maximum downward deflection $\delta$ of the beam under a set of conditions as shown in Fig. 4.12.

$$\delta = \frac{FL^3}{48EI}$$

The deflection $\delta$ depends on the input load to this component and other properties of the form. The force applied is represented by $F$. In this model, $L$ is the length of the beam. The term $E$ is elastic modulus which is a property of the material from which the beam is made. In this equation, $I$ is the second moment of area of the beam. Variable $I$ depends on the geometry and dimensions of the beam cross section.

For example, assume the beam is a standard S3 × 5.7 I-beam made of structural steel. The beam is 12 feet long. A force of 2000 pounds (8896 N) is applied at the mid-span. The desired deflection from the unloaded condition is 0.001 inch or less (0.0254 mm).

For structural steel the elastic modulus $E$ is $29 \times 10^6$ lbs/in$^2$ (200 GPa). For an S3 × 5.7 I-beam the value of $I$ is 2.52 in$^4$ (105 cm$^4$). With the parameters defined the equation for deflection can be evaluated.

$$\delta = \frac{FL^3}{48EI} = \frac{(2000 \text{ lbs}) \times \left(12\,ft \times \frac{12\,\text{in}}{1\,\text{ft}}\right)^3}{48 \times \left(29 \times 10^6 \frac{\text{lbf}}{\text{in}^2}\right) \times \left(2.52\,\text{in}^4\right)} = 0.000985 \text{ in}$$

$$\delta = 0.000985 < 0.001 \text{ in}$$

The analysis indicates that the expected deflection is less than the maximum allowed.

The form property of deflection is a function of the component input force and other form properties. If the deflection did not meet requirements, this mathematical model can be used to vary adjustable parameters to achieve a specified deflection.

The equation for deflection of the beam can be interpreted to provide insight on how aspects of the component form and input influence deflection. The deflection equation was given as:

$$\delta = \frac{FL^3}{48EI}$$

Deflection increases with increasing $\delta$ applied force $F$. Deflection is directly proportional to force, so doubling the applied force doubles the deflection. Deflection also
increases with beam length $L$. For the same load longer beams will deflect more. Deflection is particularly influenced by $L$. Deflection depends on length cubed. So, if length is
doubled, deflection increases by a factor of 8.

The material property of $E$, the Elastic Modulus, appears in the denominator. Materials
with large values of $E$ are less elastic and will bend less. The parameter $I$ is the second
moment of area and depends on the geometry of the beam cross section. Generally, the
more material that is away from the bending axis, the greater the value of $I$. I-beams have
high values of I for their weight because the beam bending axis is through the middle
portion of the beam but I-beams have a significant amount of material in the top and
bottom of the "I" away from the bending axis.

## 4.5    Component Mathematical Model Development

Mathematical models or equations that describe component behavior are often based on
the underlying physical processes or phenomena utilized in the component. In this situation, the physical phenomenon employed are expressed in terms of a mathematical
relationship or models. An expression for the behavior of the component is developed
by combining and manipulating the mathematical models of the underlying physical
processes.

The transformation of inputs to outputs by a component utilize the underlying physical
phenomena occurring in that component. If a mathematical model exists that describes the
relevant phenomena, these models can be applied to analysis of the component situation.
The mathematical equation describing the behavior of a component can be developed
from analysis of the underlying physical effects taking place in the component.

Example: Electric Generator

An electric generator is an example in which a general mathematical model of a
phenomena is applied to a specific instance in a particular component.

One type of electric generator is based on the phenomenon of electromagnetic induction. Electromagnetic induction is a fundamental property of nature. When the amount of
magnetic flux in a conductor changes, an electromotive force is induced in the conductor.
This general fundamental principle is expressed mathematically as:

$$E = -\frac{d\Phi}{dt}$$

where E is the electromotive force, $\Phi$ is the quantity of magnetic flux going through the
conductor, and $t$ is time. The induced electromotive force is proportional to the rate of

change of flux. This is a general statement of the principle of electromagnetic induction. It can be applied to any specific situation.

One specific implementation is in the form of an electric generator. Consider the form of one type of electric generator as was shown in Fig. 4.9. Coils of wire are located between magnetic poles. The coils are rotated. As they rotate the amount of magnetic flux going through the coils changes, this induces an electromotive force or voltage. This electromotive force is connected to an external circuit and provides a source of electric power.

The general mathematical expression for the phenomenon of electromagnetic induction can be applied to the specific geometry of a generator to determine the voltage that can be produced for a specific component design. For the sake of brevity, a simpler version of the generator will be analyzed. This will illustrate the application of the general principle while limiting mathematical detail.

A simplified basic generator is shown in Fig. 4.13. This represents an end view of the generator facing the end of the shaft and the side of the coils. The form of the generator is described by the number of coils, the area of the coils, and the strength of the magnetic field. The rate of rotation describes an input. The magnet field goes from the North pole of the magnet to the South pole. Some of this flux goes through the coil. The amount of flux depends on the coil size and orientation of the coil with respect to the magnetic field.

Assume that the generator has $N$ coils of wire with an area $A$. The coils are being rotated at a rate of RPM revolutions per minute. At a particular instant angle $\theta$ defines the angle between the coil and the direction of the uniform magnetic field $B$.

To determine the voltage generated, the rate of change of the magnetic flux through the coil must be determined. The amount of flux through the coil at any instant is given by:

$$\Phi = NAB \sin \theta$$

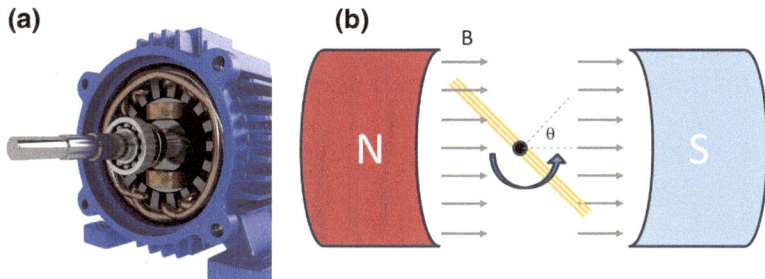

**Fig. 4.13**  Basic electromagnetic induction generator: **a** (olegbush/Shutterstock.com); **b** illustration by author

The angle $\theta$ changes with time as the coils are rotated. At a time $t$ the angle $\theta$ can be found from:

$$\theta = \omega t$$

where $\omega$ is the angular velocity in radians per second. The angular velocity can be determined from the revolutions per minute, RPM, by the equation:

$$\omega = \frac{2\pi}{60} RPM$$

Substituting for $\theta$ the rate of change of the flux can be found from:

$$\Phi = NAB \sin(\omega t)$$

$$\frac{d}{dt}\Phi = \frac{d}{dt}[NAB \sin(\omega t)] = \omega NABCOS(WT)$$

Finally, the electromotive force, or voltage, produced by the generator can be determined:

$$E = -\omega NAB \cos(\omega t)$$

The result expresses the output produced by the generator of particular form parameters and input conditions. The output depends on input as represented by the angular speed of rotation, and the form properties of coil area, number of coils, and magnetic field strength. The electric generator provides an example of how a mathematical model of component behavior is developed by applying a mathematical model of the underlying phenomenon used in the component. A basic phenomenon expressed mathematically in general form can be used to determine the output of a component based on inputs and specific form characteristics.

## 4.6    Mathematical Model Limitations

### 4.6.1    Limitations Inherent in Approximating Physical Reality

Most mathematical models are approximations of the complex physical world. Successful mathematical models depend on identification of the most significant factors influencing the behavior of a physical system and exclusion of the less significant factors. The less significant factors complicate the model without a proportionate increase in accuracy or usefulness.

Component mathematical models include assumptions, limitations, or simplifications regarding component behavior. The assumptions may be based on a determination of the most and least influential phenomena affecting component behavior relevant to its function in a particular system. The model is accurate to the extent that the assumptions are valid. The assumptions may also concern ranges of conditions in which the component is expected to operate. A key issue is whether or not the models are sufficiently accurate for the purposes for which they are intended.

The electric motor provides an example of a mathematical model developed by selecting the factors that are the most significant effects on the component's behavior. An electric motor converts electrical energy into energy of motion. A view of a motor that might be found in a home appliance or small power tool is shown in Fig. 4.14. Also shown is a simplified view of a basic motor in which the component parts are more visible.

An equation describing the output mechanical power from a basic direct current motor is given by:

$$P_M = KI\omega$$

In this equation $P_M$ is the output mechanical power. The angular speed of rotation is $\omega$. The parameter $K$ depends on the form properties of the motor.

$$K = \frac{N_C N_P}{2\pi A}\phi$$

**(a)**                    **(b)**

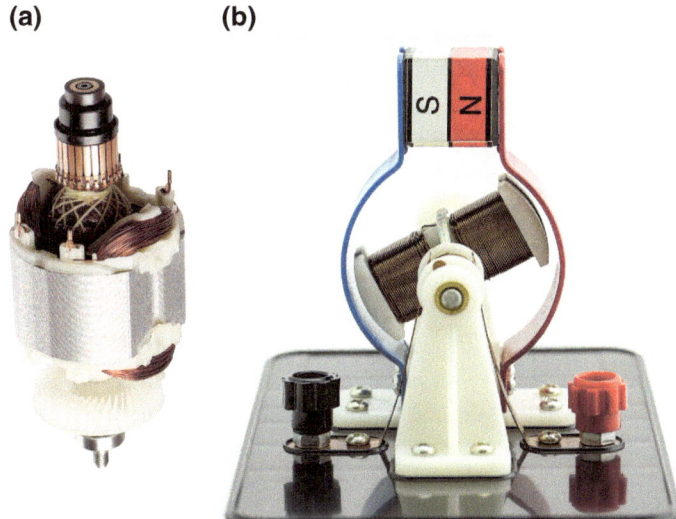

**Fig. 4.14** Simple electric motor: **a** (iStock.com/Dmytro Synelnychenko); **b** (iStock.com/William_ Potter)

$N_C$    Number of loops of conductor.
$N_P$    Number of magnetic poles.
$\phi$    Magnetic field strength per pole.
$A$      Area of conductor loop.

The development of this mathematical model assumed that the friction in the rotating parts of the motor is negligible. This is a reasonable assumption in many cases and makes it possible to develop a relatively uncomplicated expression for the power delivered by the motor. If the amount of friction present is significant compared to the magnitude of the delivered torque, then this model describing the component behavior may become inaccurate.

Motors may have minimal friction when new but can develop friction during operation. Over time the motor shaft may wear out possibly increasing friction. The degradation of friction-reducing element such as bearings and lubrication may occur. The motor may be operating in conditions that might lead to increased friction through corrosion, dust, or dirt.

### 4.6.2    Limit of Phenomena on Which Component Transformations Are Based

The phenomena utilized in the component to achieve the component function may only occur over a particular range of conditions. Outside of this range the component does not function as intended. A component mathematical model based on a particular physical principle is invalid if the assumed underlying phenomena cease to be active in the component. The mathematical model for the component cannot be used when parameters of component operation are beyond the applicable range.

*Example: Airfoil wing*
The basic wing lift formula is an example of a component equation that is accurate only for a limited range of conditions. The wing is a critical element in the generation of lift for modern aircraft. The wing lift formula quickly becomes invalid when the wing enters conditions in which the underlying phenomena changes abruptly.

A key component of the airplane is the airfoil wing. The airfoil wing has a cross sectional shape that produces an upward force or lift with a relatively small frictional force or drag. The airplane depends on generating significant lift without very much drag. For the plane to remain in the air the upward lift force must be sufficient to counteract the downward gravitational force of the plane's weight. Figure 4.15 depicts a side view of an airplane wing and the basic airfoil wing shape.

A basic mathematical model for the wing lift force $F_L$ is given by:

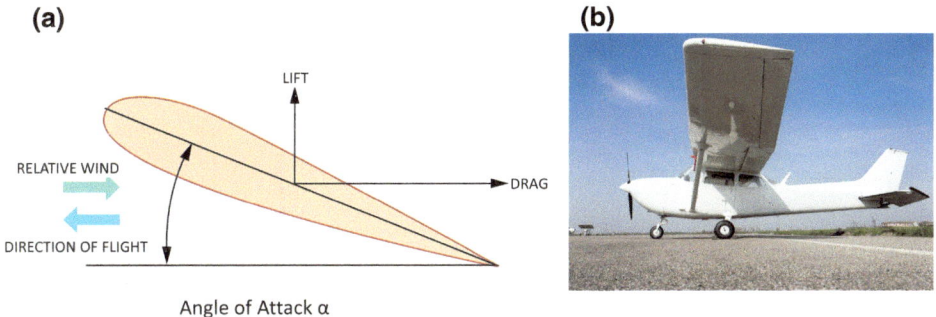

**Fig. 4.15** Basic airfoil shape and airplane wing: **a** (iStock.com/blueringmedia); **b** (iStock.com/ale
xandrumagurean)

$$F_L = \frac{1}{2}\rho V^2 A C_L$$

The terms in lift equation are:

$F_L$   Lift force on the wing.
$\rho$    Density of the air.
$V$    Relative speed of wing and air.
$A$    Wing planform area.
$C_L$   Lift coefficient.

From the mathematical model it can be seen that the lift generated by the wing depends
on the density $\rho$ of the air in which the wing is moving, the velocity $V$ of the wing
relative to the air, the wing surface area $A$. Area is defined as the planform area or the
area seen looking down on the top of the wing.

The terms in the mathematical model provide some insight into the factors influencing
the upward lift force on the wing. The combination $\frac{1}{2}\rho\ V^2$ is the kinetic energy per unit
volume of the air relative to the wing. Greater kinetic energy produces more lift.

The lift force is directly proportional to wing area $A$. Larger wings produce more lift.

The wing lift also depends on the lift coefficient $C_L$. The lift coefficient depends on the
specific geometry of the wing. The lift coefficient depends on the angle of the wing with
respect to the oncoming air stream. This angle is called the angle of attack $\alpha$. Figure 4.16
is a graph of a typical lift coefficient of a basic airfoil wing as a function of angle of
attack.

As can be seen in Fig. 4.16, initially the lift coefficient increases as angle of attack
increases. A greater wing tilt results in a larger lift force. However, near an angle of
$20°$ the lift coefficient reaches a peak. The graph shows a decrease after 20 and no lift
coefficient values are given after $21°$.

**Fig. 4.16** Typical lift
coefficient as a function of
angle of attack for basic airfoil
wing

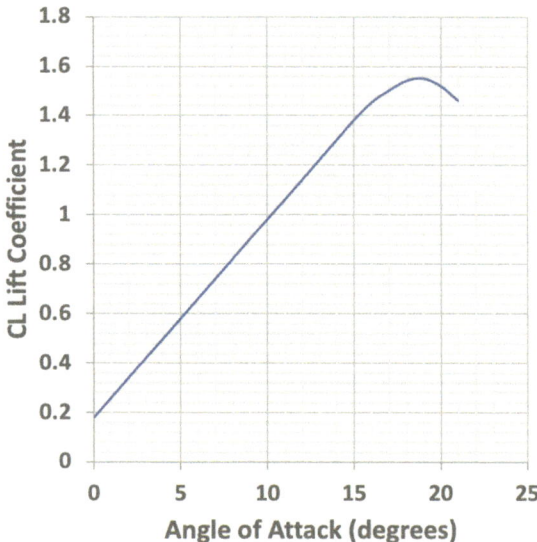

The simple expression for wing lift is subject to a limited range of operation. The formula becomes invalid as the angle of attack increases beyond about 20°. This is because the wing tilted beyond a critical angle will stop generating lift and reach a stall condition.

The lift equation approximates the lift force so long as the pattern of flow of incoming air over the wing is relatively smooth and the turbulent separated airflow over the wing is small. Smooth flow at low angles of attach is shown in the top panel of Fig. 4.17.

Lift increases as the wing is tilted at a greater angle to the oncoming air. However, if the angle of attack exceeds about 20° that pattern of flow changes abruptly. The underlying physical phenomenon affecting the wing changes. The air no longer flows smoothly over the wing but rather creates a turbulent low-pressure region on the back or downstream side of the wing. This condition is seen in the lower portion of Fig. 4.17. This type of flow results in a sudden considerable increase in drag and a loss of lift force.

Figure 4.18 is a photograph of smoke trails used to visualize flow over an airfoil wing in a wind tunnel test. The image shows an airfoil wing in a stall condition. The smoke trails do not cling to the wing but rather create a low-pressure wake that increases drag and reduces lift.

The stall angle of attack is of considerable practical importance. The stall angle limits how steeply an aircraft can ascend. If the plane tilts upward at an angle exceeding the stall angle, the lift force supporting the plane will suddenly decrease resulting in possible crash conditions.

The basic wing lift equation describes the lift force generated by a wing as a function of air speed, air density, wing area, and the lift coefficient that depends on the specific geometry of the wing and the angle of the wing relative to the oncoming air. The equation is valid so long as air is flowing smoothly over the wing. When the angle of attack

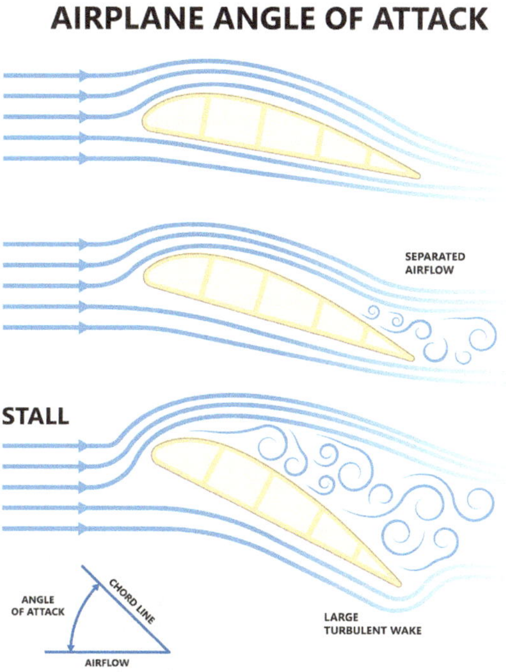

**Fig. 4.17** Visualization of flow over a wing at increasing angle of attack (Pepermpron/Shutterstock.com)

**AIRPLANE ANGLE OF ATTACK**

SEPARATED
AIRFLOW

STALL

ANGLE
OF ATTACK
CHORD LINE

AIRFLOW

LARGE
TURBULENT WAKE

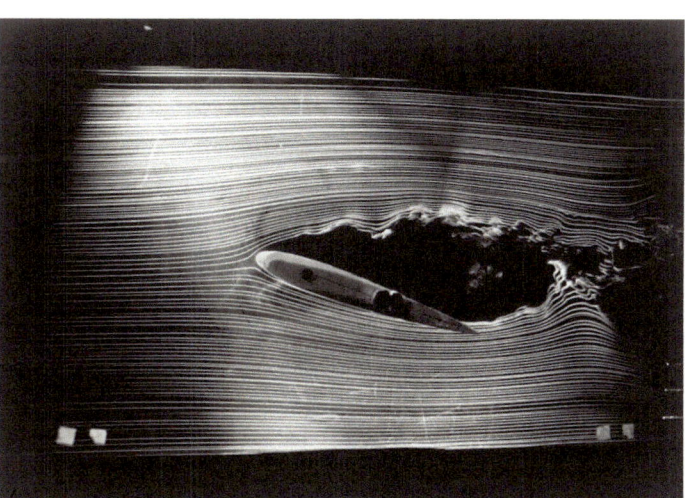

**Fig. 4.18**   Flow over a wing approaching a stall condition [photo reversed]. (Available from Defense Visual Information Distribution Service. NASA Identifier: L89-10604. Public Domain)

is increased beyond a critical value, the flow pattern of the air over the wing changes abruptly. The mathematical model becomes invalid because the underlying physical processes affecting the wing have changed. The mathematical model used to describe wing behavior is no longer appropriate for the conditions.

Examples

Provided that the angle of attack is less than the stall angle, the lift equation is a useful mathematical model in determining upward lift force as a function of other parameters. Some examples will illustrate the applicability of the lift equation for different types of aircraft.

*Example: Boeing 777*

As an example, consider the Boeing 777. The 777 is one of the largest commercial airliners with a capacity of approximately 350 passengers. A Boeing 777 is shown in Fig. 4.19.

The lift force on a 777 can be calculated for typical take off conditions. The relevant parameters have the following values:

$A$     Wing Area $= 420$ m$^2$
$C_L$   Lift Coefficient for 777 Wings at Takeoff $= 3.0$
$V$     Take Off Speed $= 170$ mph $= 76$ ms$^{-1}$
$\rho$   Density of Air at Ground Level $= 1.23$ kg/m$^3$

**Fig. 4.19**  Boeing 777 (iStock.com/Alvin Man)

$$F_L = \frac{1}{2}\rho V^2 A C_L = \frac{1}{2}(1.23\,\text{kgm}^{-3})(76\,\text{ms}^{-1})^2(420\,\text{m}^2)(3.0)$$

$$F_L = 2.936 \times 10^6\,\text{N} = 660{,}000\,\text{pounds}$$

The wings of the Boeing 777 generate about 660,000 pounds of lift force (2.93 MN) during takeoff. This means that the total weight of the plane including passengers, cargo, and fuel can not exceed 660,000 pounds for the plane to be able to leave the ground. This result is comparable to the reported 777 maximum design takeoff weight of 508,000–777,000 pounds.

*Example: Ultralight Aircraft*
An ultralight aircraft is a small single-passenger plane. Currently in the United States an ultralight aircraft is defined as a vehicle with an empty weight of less than 254 pounds (115 kg) and a maximum speed of 64 mph (102 km/h). Under these conditions a pilot's license is not required to operate but operation is restricted to flights over unpopulated areas.

The lift force on a typical ultralight can be calculated for typical take off conditions. The relevant parameters have the following values:

$A$     Wing Area $= 12$ m$^2$
$C_L$   Lift Coefficient at Takeoff $= 1.2$
$V$     Take Off Speed $= 40$ mph $= 17.9$ m/s.
$\rho$   Density of Air at Ground Level $= 1.23$ kg/m$^3$

$$F_L = \frac{1}{2}\rho V^2 A C_L = \frac{1}{2}(1.23\,\text{kgm}^{-3})(17.9\,\text{ms}^{-1})^2(12\,\text{m}^2)(1.2)$$

$$F_L = 2135\,\text{N} = 480\,\text{pounds}$$

The typical ultralight is able to produce about 480 pounds of lift force (2.13 kN) during takeoff. If the plane is close to the legal limit of 254 pounds empty, this means that the total weight of the plane, including pilot and fuel, cannot add more than an additional 226 pounds for the plane to be able to leave the ground (Fig. 4.20).

## 4.6.3 Improving Component Mathematical Models

Mathematical models of component behavior that are based on models of underlying physical phenomena are limited by the extent to which the modeled phenomena determine component behavior. Limitations occur if other physical processes take place during

**Fig. 4.20**   Ultralight aircraft (iStock.com/Nnehring)

aspects of component operation that are not included in the model. Clearly, mathematical models can be improved if additional physical effects are included. Typically, this causes the model to become more mathematically complicated and potentially more difficult to use. Inherent in the use of mathematical models is a tradeoff between convenience and sufficient accuracy needed to appropriately inform a specific decision.

The mathematical model for lift on a wing is an example of a model in which accuracy can be improved by including additional physical phenomena. The basic lift equation does not take into account the details of wing shape or the effects occurring at the end of the wing tips.

At the wing tips high pressure air under the wing moves around the tip to the top of the wing. Since the wing is moving forward all the time that the plane is in motion, this swirling of air from the underside of the wing results in a spiraling vortex steaming continuously behind the airplane. Under some atmospheric conditions, these trailing vortices are visible as shown in Fig. 4.21. Note that trailing vortices from the wings are a different effect from exhaust trails sometimes visible from jet engines.

The effect of trailing vortices is a reduction in wing lift. This effect is modeled as a reduction in wing angle of attack. The effect reduces lift in the same way that a smaller attack angle results in lower lift coefficient.

A mathematical model for the effect of trailing vortices is given by:

$$\Delta\alpha \approx \frac{C_L}{\pi ar}$$

$$ar = \frac{b^2}{A}$$

**Fig. 4.21** Aircraft with trailing vortices from wing edges [cropped] (iStock.com/Jetlinerimages)

where:

$\Delta\alpha$    Approximate effective reduction in angle of attack.
$C_L$    Lift coefficient assuming no effect.
*ar*    Aspect ratio of the wing.
*b*    Wing span.
*A*    Wing planform area.

While the effect is relatively small, many modern aircraft incorporate upward facing winglets on the wing tips to reduce induced vortices.

## 4.7    System Models From Component Models

A component can be combined with other components to form a system. In a similar fashion, the mathematical models of components can, in some cases, be combined to create a mathematical model of the resulting system.

In a basic combination of components, an output of one component becomes an input to another component. The mathematical models of each component should then share at least one common term. In a simplest case the component models can be combined by replacing the expression for the output of one component into the input of the other component.

This section will demonstrate by a limited example the combining of mathematical models of components into a larger model describing the resulting system. In the process some of the other main topics of this chapter will be reviewed. The function and behavior of the components is based on underlying physical phenomena implemented through component form. Mathematical models of component behavior can be created by applying expressions of general principles to the specific component form. The mathematical models involve simplifications to limit complexity, but are sufficiently accurate to provide useful insights about system behavior.

*Example: Domestic Refrigerator*

Consider a typical domestic refrigerator as is shown in Fig. 4.22. The domestic refrigerator provides an example of combining component models. A key effect utilized in the refrigerator is the energy absorbed by a substance changing from a liquid to a gas. In the refrigerator, the circulating refrigerant liquid absorbs heat from the interior space and transfers this heat to the exterior room.

Consider a diagram of a portion of the refrigerator shown in Fig. 4.23. The air inside the refrigerator is cold. Heat leaks into the refrigerator through the outside wall. The cold inside air interacts with the enclosure wall and gains heat. If not removed, the continuous addition of heat to the inside air would warm up the interior of the refrigerator.

The air comes in contact with the evaporator component inside the refrigerator. In the evaporator the excess heat energy in the air is transferred to the refrigerant. The air

**Fig. 4.22**  Typical domestic refrigerator (iStock.com/ photo_world)

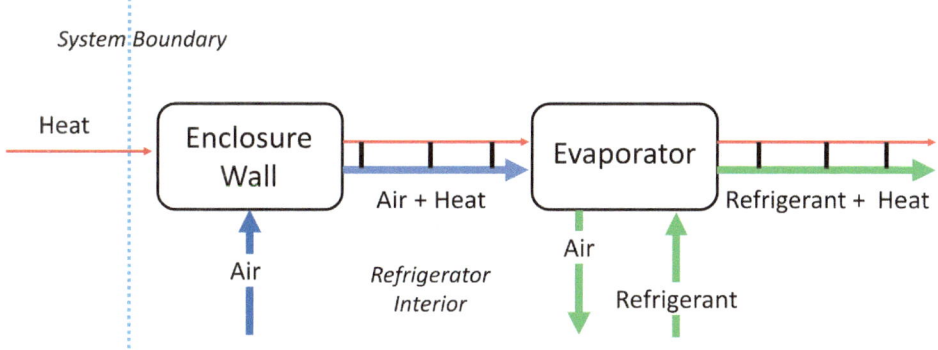

**Fig. 4.23** System diagram of a portion of the refrigerator system

leaves the evaporator cooled down and circulates back to the refrigerator interior. The heat removed from the air is gained by the refrigerant. The refrigerant with the added heat leaves the evaporator to release the heat to the outside of the system.

*Refrigerator Evaporator*

The primary effect underlying the operation of a refrigerator is the energy required to create a change in state from a liquid to a gas. This takes place in a component called the evaporator. Figure 4.24 shows the location of the evaporator in a typical refrigerator.

Figure 4.25 is a photograph of an evaporator inside of a refrigerator. The evaporator is simply a serpentine tube inside of which a refrigerant liquid flows.

**Fig. 4.24** Refrigerator showing evaporator (Colin Hayes/Shutterstock.com)

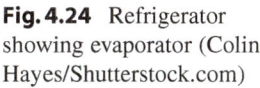

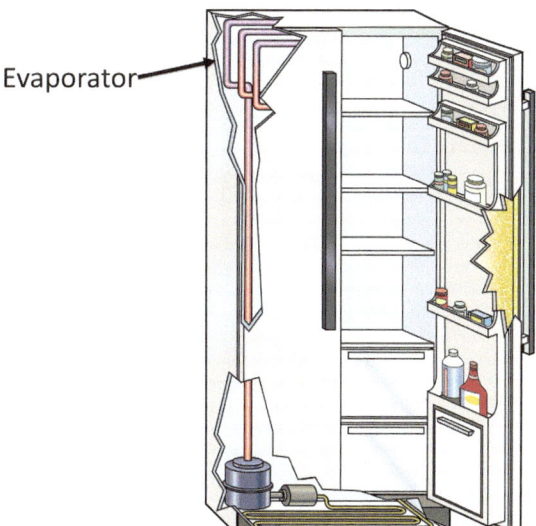

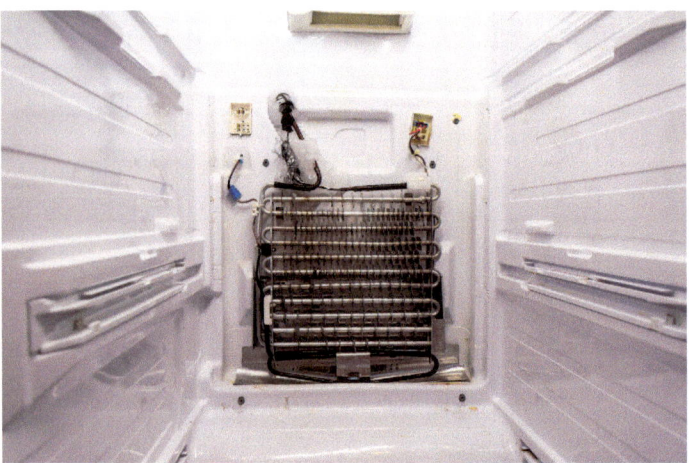

**Fig. 4.25** Photograph of refrigerator evaporator piping inside refrigerator (iStock.com/Michael Vi)

The evaporator is the coldest part of the refrigerator and keeps the refrigerator contents cold. Heat removed from the refrigerator interior space causes a change in state of the refrigerant from the liquid to the gas phase. The heat energy removed from items inside the refrigerator is gained by the refrigerant liquid. This causes the refrigerant to change phase from a liquid to a gas. Figure 4.26 illustrates this process.

To evaporate and change state from a liquid to a gas, energy must be absorbed by the liquid. The energy breaks the bonds keeping the liquid molecules together, liberating them to then move about freely in the gas phase. The amount of heat energy needed to transform the liquid into a gas is called the heat of vaporization $L_V$. The heat of vaporization is a characteristic physical property of a material. For a particular material a specific amount of heat is needed to transform a precise amount of mass from the liquid to the gas phase. The heat of vaporization is measured in units of heat per mass. Heat of vaporization is a property that describes one aspect of the form of the refrigerant.

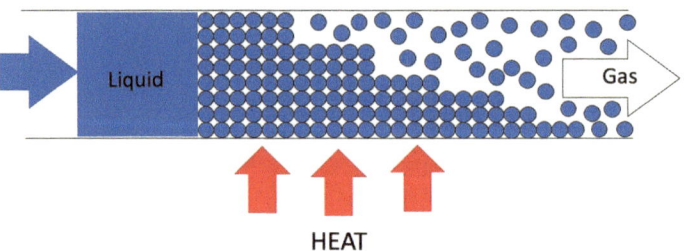

**Fig. 4.26** Principle of evaporation (illustration by author)

It is possible to develop a mathematical model in the form of an equation describing the evaporator based on underlying effects and parameters that describe the situation.

The goal is to determine the amount of liquid that must flow through the evaporator for a particular amount of heat input.

$L_v$ = Heat of vaporization of the liquid (J/kg).

$\dot{m}$ = Flow rate of refrigerant liquid through the evaporator (kg/s).

$\dot{Q}$ = Rate of heat removal from the items in the refrigerator (J/s).

$L_v$ is the heat of vaporization of the liquid. This is measured in units of energy per mass. The flow rate of refrigerant liquid through the evaporator is $\dot{m}$. Flow is measured in mass per time. $\dot{Q}$ is the amount of heat removed from the items in the refrigerator. The dimensions of heat flow are energy per time.

The amount of heat absorbed is equal to the heat needed to vaporize a unit mass multiplied by how much mass is flowing through the evaporator:

$$\dot{Q} = \dot{m} L_V$$

Focusing on the physical units of each term clarifies this result:

$$\frac{J}{s} = \frac{kg}{s} \times \frac{J}{kg}$$

In this case a mathematical model has been developed based on the fundamental physical processes taking place in this refrigerator component. The fundamental physical process at work is the transformation from a liquid to a gas. This is characterized by the heat of vaporization, a physical property of the material. How much material vaporizes in a given amount of time determines the amount of heat absorbed in that time.

*Refrigerator Enclosure*
A function of the refrigerator enclosure is to reduce the flow of heat from the room into the inside of the refrigerator. The underlying phenomenon of interest here is the flow of heat from a higher to a lower temperature. A mathematical model for heat flow into the refrigerator can be obtained by applying the general principle of heat flow.

A general mathematical model for the flow of heat through solids is Fourier's Law of Conduction. Consider a solid material such as a wall shown in Fig. 4.27 with different temperatures on each side. Assume that $T_2$ is greater than $T_1$. Heat will flow from the hotter to the colder side through the solid material.

Fourier's mathematical model of heat conduction from the hotter to the colder side is:

$$\dot{Q} = kA \left( \frac{T_2 - T_1}{\Delta x} \right)$$

The terms in the heat conduction equation are:

$\dot{Q}$     Heat flow from higher to lower temperature (J/s).

**Fig. 4.27**  Heat flow due to
conduction (illustration by
author)

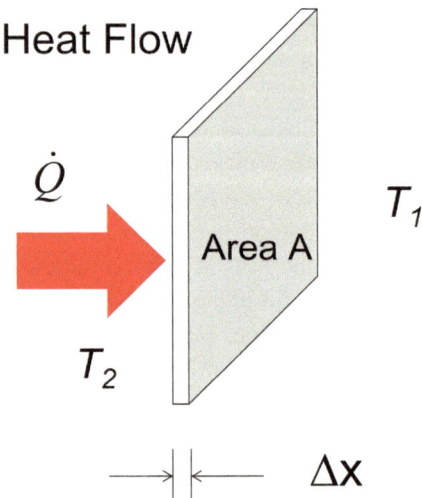

*k*      Thermal conductivity, a property of the material (W/mC).
*A*      Area through which heat flows (m$^2$).
$\Delta x$    Thickness of material (m).

The general mathematical model for the phenomenon of heat conduction can be applied
to the specific form and parameters of the refrigerator. The goal is to illustrate the nature
of component mathematical models so the approach here will be to carry out a simplified
analysis that still includes the major effects.

   With the doors closed, the refrigerator can be viewed as an insulated box. The hinges
and seams will be ignored. Assume that the enclosure wall has a uniform thickness on all
sides. In most cases, door openings and closings are actually relatively infrequent. Nearly
all of the time the refrigerator is just sitting in a warm room trying to keep itself cold.
The main heat flow into the refrigerator then is the heat conducted in through the walls.

   The general Fourier's Law of Conduction can be applied to the specific instance of
the refrigerator. In this case $\Delta x$ is the thickness of the enclosure walls, *k* is the thermal
conductivity of the enclosure wall material. The inside of the refrigerator is colder than
the room, so $T_1$ is the internal temperature and $T_2$ is room temperature. The area *A*
through which heat flows is the entire surface area of the refrigerator exposed to room
temperature $T_2$. The entire surface area of the refrigerator consists of the front, back, top,
bottom, left and right sides. Figure 4.28 is a representation of the refrigerator with the
width, depth, and height labeled.

   The total surface area is given by:

$$A = 2 \times (w \times h) + 2 \times (w \times d) + 2 \times (h \times d) = 2(wh + wd + hd)$$

**Fig. 4.28** Representation of the refrigerator as a simplified box (illustration by author)

The heat flow $\dot{Q}$ into the refrigerator can be expressed as:

$$\dot{Q} = kA\left(\frac{T_2 - T_1}{\Delta x}\right) = 2(wh + wd + hd)k\left(\frac{T_2 - T_1}{\Delta x}\right)$$

The result is a mathematical model for the heat input into the component as a function of the form properties and conditions on the system.

$$\dot{Q}_{Enclosure} = \dot{Q}_{Evaporator}$$

$$kA\left(\frac{T_2 - T_1}{\Delta x}\right) = \dot{m} \cdot L_V$$

$$\dot{m} = \frac{kA}{L_V}\left(\frac{T_2 - T_1}{\Delta x}\right)$$

The mathematical model of system behavior obtained from models of the components provides insights into how relevant factors influence the system. Generally, the results are consistent with what might be expected from a qualitative understanding of the underlying phenomenon. However, the mathematical model enables quantitative predictions that can be used in the design process to help ensure that the resulting system will meet specific quantitative requirements.

The difference between the room temperature $T_2$ and the refrigerator interior temperature $T_1$ appears in the numerator of the mathematical model equation. As $T_2$–$T_1$

increases, the mass flow rate increases. This appears reasonable; a warmer room or colder refrigerator will require greater flow of coolant.

The refrigerator exterior surface A is also in the numerator of the flowrate equation. As area $A$ increases, the mass flow rate increases. Larger size refrigerators will require greater flow of coolant.

The flowrate is directly dependent on $k$, the thermal conductivity of refrigerator enclosure wall insulation. As $k$ increases, the mass flow rate increases. Insulation that has higher conductivity will result in a refrigerator needing greater flow of coolant.

The refrigerator enclosure wall thickness $\Delta x$ is in the denominator of the mathematical model equation. Therefore as $\Delta x$ increases, the mass flow rate decreases. Coolant flow rate is inversely proportional to wall insulation thickness. This appears consistent with what might be expected. A refrigerator with thicker walls and insulation will result in a lower flow of coolant.

The flow rate of coolant to sustain a low refrigerator temperature is also inversely proportional to $L_v$, the heat of vaporization of refrigerant. As $L_v$ increases, the mass flow rate decreases. Refrigerant that absorbs more heat while evaporating will result in the refrigerator needing a lower flow of coolant.

The example of the refrigerator illustrates major aspects of technological system components. The component function depends on application of physical phenomena in the context of the specific component form. Mathematical models can be developed to describe component behavior. These models can be used to help ensure that components in a system achieve quantitative specifications on system form and function. Mathematical models are useful in demonstrating how component behavior depends on form parameters. Model accuracy can be limited by the extent to which the included effects dominate component behavior. As technological systems are created by combining components, some technological system mathematical models can be developed through combinations of constituent component models.

## Bibliography

Anderson, John David. *Introduction to Flight*. McGraw-Hill Higher Education, 2005.

Beer, Ferdinand Pierre, Elwood Russell Johnston, and John T. DeWolf. *Mechanics of Materials*. McGraw-Hill Higher Education, 2006.

Chapman, Stephen J. *Electric Machinery Fundamentals*. McGraw-Hill Education, 2005.

Halliday, David, Robert Resnick, and Jearl Walker. *Fundamentals of Physics*. Extended 7th edition. Hoboken, NJ: Wiley, 2004.

McCabe, Warren, Julian Smith, and Peter Harriott. *Unit Operations of Chemical Engineering*. McGraw-Hill Education, 2005.

Reisel, John R. *Principles of Engineering Thermodynamics*. 1st edition. Boston, MA: Cengage Learning, 2015.

Stoecker, Wilbert F., and Jerold W. Jones. *Refrigeration and Air Conditioning*. McGraw-Hill, 1986.

# System Characteristics in a Technological System Context

**5**

## 5.1 Chapter Overview

- This chapter focuses on some distinctive characteristics of systems. How these system features appear in technological systems is emphasized.
- Systems exhibit characteristic behaviors. Technological systems demonstrate these behaviors in distinctive ways. System behavior is emergent.
- The action of technological systems emerges from the interactions of the system components.
- Unanticipated emergent system behaviors are possible due to unexpected effects and interactions. Systems are hierarchical.
- A technological system can be viewed as a set of sub-systems. What is viewed as a system in one context might be considered a component in another circumstance.
- Systems are dynamic. The transformation of the inputs into the outputs when the system is operating demonstrates the dynamic nature of technological systems.
- Systems frequently contain feedback or flows loops within the system that can influence system dynamic behavior. Technological systems frequently employ feedback often in the automatic control of system operation.
- Systems including technological systems of interacting components can exhibit nonlinear behavior such that a change in output occurs that is not in simple proportion to changes in inputs.
- Some systems have a complex internal component structure, multimodal operation, and multifaceted interactions and dependences on other systems.
- Characterizations and models of such systems will necessarily be a simplification based on the most significant elements and consequential interactions.

© The Author(s), under exclusive license to Springer Nature Switzerland AG 2024
J. Krupczak, Jr., *Understanding Technological Systems*, Synthesis Lectures
on Engineering, Science, and Technology, https://doi.org/10.1007/978-3-031-45441-7_5

Systems are a group of interacting objects, forming a network, achieving a common outcome. Systems as a phenomenon themselves display some behaviors viewed as characteristic of systems. These include emergence, hierarchies, dynamic behavior, feedback loops, homeostasis, and non-linearity. Many different types of systems exist such as biological, environmental, political, economic, managerial, and cultural. Characteristic system behaviors will appear somewhat differently in different types of systems. This section describes how some of the distinctive characteristics of systems appear in the context of technological systems.

## 5.2  Emergence

Emergence expresses the idea that the behavior of the system emerges from the interaction of its constituent elements with each other and the external environment. Emergence is a fundamental characteristic of systems. The system elements by their interactions produce an outcome that is not possible in absence of the interactions. Synergy, or the whole is greater than the sum of its parts expresses the essential nature of systems.

For most systems created by people, the expectation of specific emergence effects is a deliberate reason for establishing the system. However unintentional or unanticipated component-to-component and component-to-environment interactions can occur in all types of systems. These unforeseen interactions have the potential to produce unexpected and possibly undesired emergent behaviors. The possibility of non-deliberate interactions occurring can increase for larger systems with more complex networks of interactions. In the study of some categories of systems, occurrence of unanticipated emergence is a topic of major interest.

For technological systems in the modern era, system functional behavior is deliberately engineered. Technological systems are created to provide an output of greater value to the user than the uncoordinated function of the individual pieces. The behavior of the technological system and ability to transform the available inputs into the desired outputs emerges from the interactions of the assembled system. The technological system's useful functions occurs when the system is interacting with the external environment. It is the interactions among the components that result in the desired outputs from the given inputs.

Unintended emergent behaviors that were not anticipated, both undesirable and desirable, can occur in systems from unexpected interactions between components and the inputs and outputs to the external environment. Several factors reduce the possibility of engineered technological systems producing unexpected emergent behaviors in terms of technological function. The systems are generally developed to meet specific quantitative requirements regarding form and function. So, for technological systems, the expected system behavior is well-defined during the development phase. In addition, the behavior of most constituent components used in technological systems is long-understood and well-characterized in terms of performance expectations, operating limits, and failure history. Mathematical models used in system design, prototype testing, and awareness of the need to avoid legal liabilities for negligence or failure to mitigate readily foreseen problems, all reduce the possibility of unanticipated system behavior after technological products are released. Nevertheless, lapses of quality control, unusual component failure modes, and novel use cases can lead to unexpected emergent behaviors often leading to product recalls or design changes.

Behaviors that occur due to the interaction of the components within a particular technological system is one aspect of emergent behavior. A separate aspect is emergent behaviors through the interaction of a technological system with other technological, environmental, and human systems. Technological systems become components in larger socio-technical and environmental systems. Technological products interact with and influence the environment and other human-created systems both technological and non-technological. This aspect of technological systems is discussed in a later chapter.

Unintended emergent behavior is a situation distinct from the functioning system being used for purposes other than the design intent. These might be called unanticipated applications of the system. Unanticipated applications and unintended interactions with other systems are addressed in a later chapter.

### 5.2.1 Example: Classic Grain-Grinding Windmill

As noted for technological systems, the system characteristic of emergence is well-recognized. System components interacting with each other and the environment external to the system convert the available inputs into desired outputs. The benefits of form manipulations and component synergy are sometimes more readily appreciated in less-familiar examples.

The classic medieval flour-grinding windmill is an example of simple materials and components enabling a significant community benefit of flour-grinding with minimal human labor. The windmill consists of components made only from stone, wood, rope, and simple hand-made iron shapes. The functioning system of the interaction of these components converts inputs of energy of motion from the wind and wheat kernel material into ground flour with control input from the operating miller. Figure 5.1 shows a classic grain-grinding windmill and some of the internal wooden gears.

### 5.2.2 Example: SpaceX Crew Dragon Anomaly

Emergent outcomes from technological systems result from component interactions. Technological systems are created based on expected component behavior and performance.

**Fig. 5.1** Classic grain-grinding windmill: **a** windmill (iStock.com/Yasonya); **b** gears in old windmill (iStock.com/ewg3D); **c** wooden gears (iStock.com/M-Production)

Component failures can result in unanticipated interactions. The unanticipated component interactions within the system can result in undesirable emergent behaviors such as degraded system performance or system failure. The SpaceX Crew Dragon 2019 anomaly provides an example of an unanticipated undesirable emergent outcome (Fig. 5.2).

The SpaceX Crew Dragon capsule has successfully transported numerous astronauts and space tourists to and from earth orbit and the International Space Station. However, on April 20, 2019, a Crew Dragon capsule exploded during testing. There were no injuries, and the capsule was yet to transport any passengers at that time, but the capsule was completely destroyed in what was called an "anomaly" or, in other words, unexpected emergent behavior of the system.

The anomaly occurred during testing of the capsule thruster subsystem. The thrusters are a series of small rocket engines that are used to make small adjustments to a capsule trajectory while in space. The small thruster engines are different from the large main engines that are used in liftoff to achieve orbit. The SpaceX Crew Dragon also uses the somewhat larger "SuperDraco Thrusters" which are used to accelerate the capsule away from the main rocket if needed.

Capsule thrusters are a well-established spacecraft subsystem. The basic design of thruster systems, shown in Fig. 5.3 dates back to at least the US Gemini program of the 1960s. The thruster rockets use monomethyl hydrazine (MMH) as the propellant fuel and nitrogen tetroxide (NTO) as oxidizer. In space, in the absence of gravity, some method must be devised to deliver the propellant and oxidizer to the thruster. Helium gas at high

**Fig. 5.2** SpaceX Crew Dragon Capsule recovered after successful mission to the International Space Station (NASA. Public Domain. https://images.nasa.gov/details/KSC-20190308-PH_SPX01_0001)

pressure is used to pressurize the propellant and NTO tanks and drive the fuel and oxidizer from their storage tanks to the thruster. In the close quarters of the capsule, various tubing is used to connect the helium tank to the NTO and propellant. For simplicity, the interconnect tubing is shown as boxes in the figure. Check valves are placed between the helium tank and the other tanks. Check valves are one-way valves that allow flow in one direction but not the other. In the basic thruster system design, the check valves are intended to allow helium to flow toward the propellant and NTO tanks but not allow flow in the reverse direction.

SpaceX investigated the cause of the explosion and published their analysis on July 15, 2019. The first aspect of the problem is illustrated in Fig. 5.4. Prior to testing, a leak allowed the NTO oxidizer to accumulate in the helium tubing on the helium side of the check valve.

The actual problem occurred when the thrusters were tested. During the thruster operation and testing, high pressure helium was released following the normal procedure to pressurize the NTO and propellant tanks. However, this burst of high-pressure helium pushed a slug of the leaked NTO into the check valve. This resulted in structural failure within the check valve. The titanium metal of the failed valve reacted with the NTO and began burning. The burning valve ignited the nearby propellant and NTO resulting in the

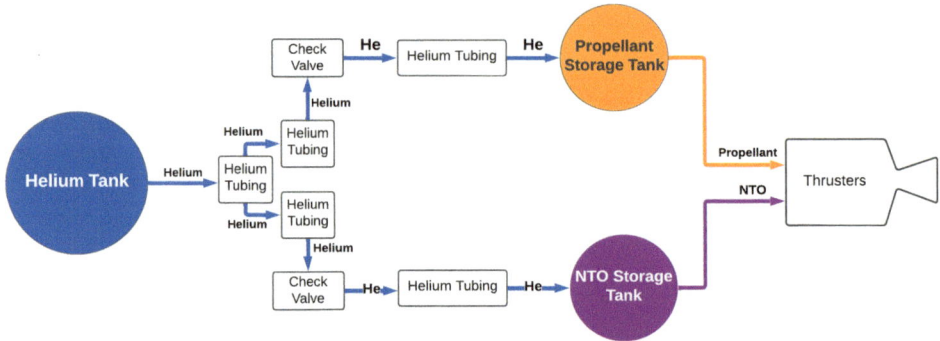

**Fig. 5.3** Basic design of a typical space capsule thruster system (diagram by author)

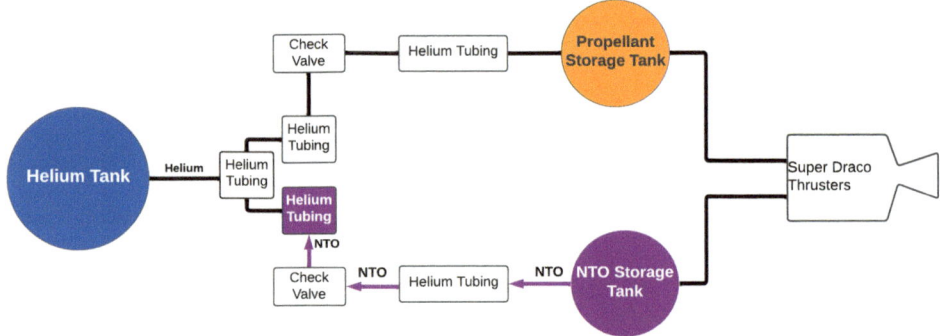

**Fig. 5.4** Leak allows NTO oxidizer to accumulate on helium side of check valve (diagram by author)

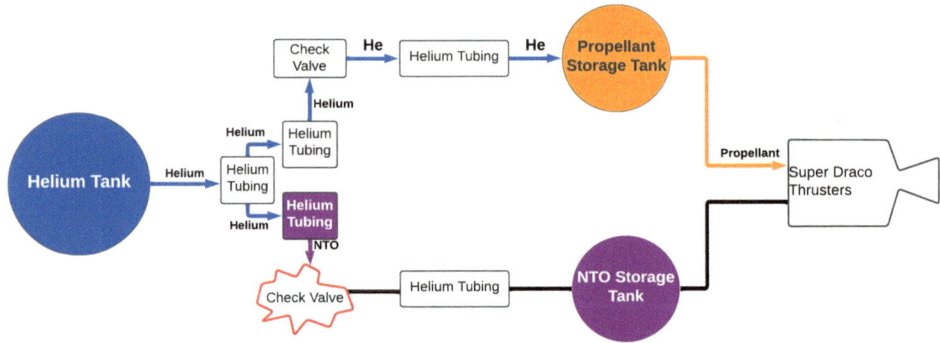

**Fig. 5.5** High pressure helium forces leaked NTO into check valve damaging valve leading to titanium valve and NTO reaction (diagram by author)

explosion that destroyed the capsule. Figure 5.5 summarizes this phase of the process. SpaceX noted that the reaction of shattered titanium from the valve with NTO was an unexpected phenomenon not previously known to SpaceX.

During the anomaly the system components were interacting resulting in emergent behavior but obviously not desired behavior. The explosion resulted from the interaction of several components. First an unexpected behavior of a component leak allowed NTO into the helium tubing. The function of tubing is normally to direct the flow of liquid, but the form of tubing can also serve as a storage volume as happened in this case. Helium gas then forced the NTO into the valve at conditions that exceeded the valves intended operating environment. Structural failure of the valve resulted in titanium and NTO interaction leading to combustion. The combusting valve interacted with the nearby propellant and NTO tanks. Without the interactions between components this situation would have been simply a leaking valve, a broken valve, and some pieces of burning titanium. Interactions resulted in the emergent behavior of the explosion.

## 5.3  Hierarchical Nature of Technological Systems

A hierarchical structure is a general property of systems. Systems are hierarchical meaning that a system can be viewed as a set of sub-systems. This might be stated as some systems are a "system of systems." Alternatively stated, systems can be composed of networks of other systems. The precise nature of the sub-system structure varies with different system types. Extensive complex systems are typically hierarchies or successive tiered layers of systems that contribute functionality building into a more complex system structure. The hierarchical nature of systems might also be described as cascades of networked subsystems.

The hierarchical perspective is a vital concept in describing the nature of technological systems. The construction of technological systems, particularly complex ones, includes multi-components subsystems necessary to achieve overall system functionality. In this way, technological systems can be viewed as a "system-of-systems."

Technological systems can be complex with thousands of elements and interactions. Such complex systems are difficult to understand as a whole. Recognizing hierarchies in technological systems is useful in managing the complexity of these systems. For a system with tens of thousands of components, attempting to design or analyze all of these components and their interactions simultaneously as a hierarchy with only one level is unlikely to be a profitable approach. The features of the system structure and the major interactions occurring are more readily discerned by identifying a system hierarchy that divides the system into smaller more manageable functional units.

### 5.3.1   Component-System Simultaneity

One way to characterize the hierarchical nature of technological systems is to adopt the perspective that there is not necessarily an absolute distinction between what is a system and what is a component. Both terms are frameworks that define how a particular entity will be treated in a given analysis. A system is considered to be made up of constituent components providing subfunctions which contribute to the overall system function. A component is viewed as an individual entity and not divided into any further detail. In a given analysis, entities treated as components are not decomposed into subfunctions. Something treated as a component in one situation can be viewed as a system in another analysis.

The refrigerator evaporator provides an example of applying either the component or system perspective. The evaporator transfers heat from the air inside the refrigerator to the refrigerant liquid. This keeps the interior of the refrigerator cool. Figure 5.6 shows an evaporator. The evaporator is an air-to-liquid heat exchanger. Thermal energy from the air is transferred to the liquid. In the component view, the evaporator transfers energy from air to liquid. Figure 5.7 illustrates the component perspective. If air and heat (thermal energy) are input, the thermal energy will be transferred to the liquid. No additional details of internal interactions are considered.

The evaporator itself can also be viewed as a system as well. This is shown in Fig. 5.8. In this case the evaporator is considered to be made up of components which combine to create the overall transformations that takes place in the system. The evaporator consists of tubes through which the coolant liquid flows. A large number of metal strips or fins

**(a)**                    **(b)**

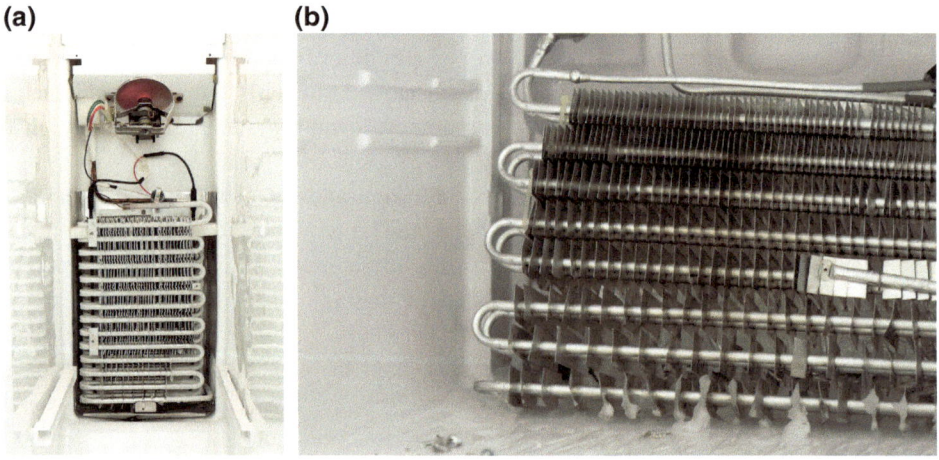

**Fig. 5.6**  Refrigerator evaporator is an air-to-liquid heat exchanger: **a** evaporator (iStock.com/Don Nichols); **b** heat exchanger (iStock.com/Pixygirlly)

**Fig. 5.7**  Evaporator treated as a component

**Fig. 5.8**  Evaporator air-to-liquid heat exchanger treated as a system (diagram by author)

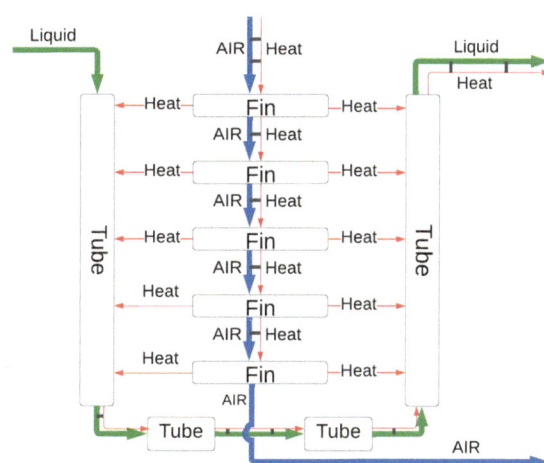

are attached to the tubes. The fins have a significant amount of surface area facilitating contact with the air. Warm air transfers heat to the metal fins. The metal fins in turn conduct the heat to the tubes where it is finally transferred to the liquid. As a system, the evaporator consists of a large number of interacting fins and tubes. Each of these components contributes a subfunction to the overall evaporator function of transferring thermal energy from the air to the liquid.

Whether treated as a system or a component the overall function of the component-system is the same. The heat exchanger transfers thermal energy from air to liquid. This function and the inputs and outputs are unchanged whether the device is considered to be a single entity or a system composed of other components. The function, inputs, and outputs are the same in each view.

## 5.3.2   Systems Can Become Components in Other Systems

Another perspective on the hierarchical nature of systems is to consider that a system can become a component in another system. A device that is treated as a system of components in one situation can also become a component in another technological system. When the system becomes a component in another system, the components of that system are no longer the focus of the analysis. The system becomes a component and is viewed in terms of its overall function along with the required inputs and resulting outputs. The details of the components of that original system are deliberately masked.

As an example, consider the centrifugal pump as shown in Fig. 5.8. This pump is used to pressurize liquids such as water so they can flow through some type of impedance or restriction. The centrifugal pump is an energy transfer device which transfers energy of motion from a rotating shaft to the fluid by means of moving blades or an impeller. When viewed as a system the major pump components include the inlet, rotating shaft, impeller, housing, and outlet (Fig. 5.9).

The centrifugal pump system becomes a component in the automobile engine as shown in Fig. 5.10. The automobile internal combustion engine converts chemical energy of the input fuel and air into kinetic energy of motion. A pump is needed to circulate coolant to help maintain the engine at an appropriate temperature.

The internal combustion engine system becomes a component in the automobile. Besides the engine, major automobile components of the automobile system include the power train, brakes, steering, and suspension. The working automobile includes every

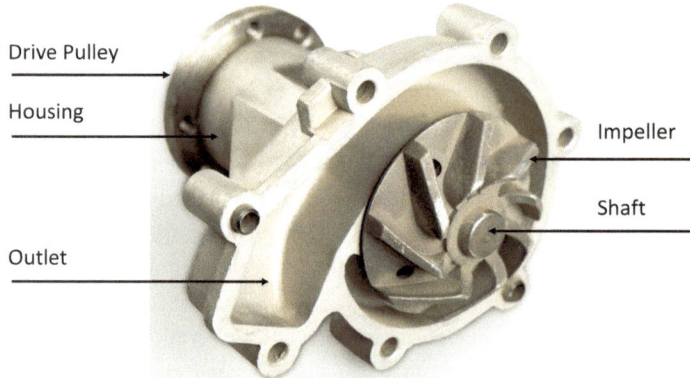

**Fig. 5.9**  Components of centrifugal pump system (iStock.com/artisteer)

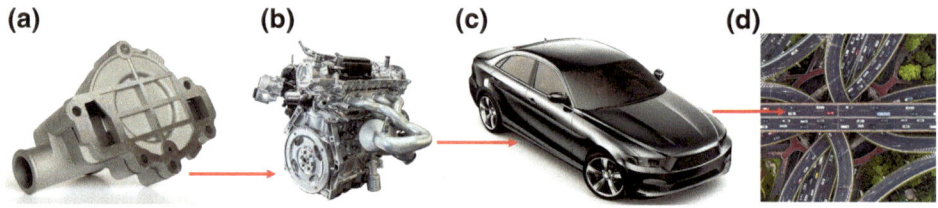

**Fig. 5.10**  Systems becoming components in other systems: **a** water pump (iStock.com/ooiphotoo); **b** automobile engine (iStock.com/luxizeng); **c** automobile (iStock.com/Vladimiroquai); **d** highway (iStock.com/JaCZhou)

individual component of the centrifugal coolant pump along with thousands of other component pieces subsumed within its subsystems.

The automobile in turn can be considered as a component in what might be called a transportation system. The transportation system includes other automobiles, buses, trucks and other vehicles. It includes interconnected roadways, toll systems, and traffic control signals. Also part of the transportation system is the maintenance and repair equipment for both the many types of vehicles and the roadways.

The process of a system becoming a component in another system is dependent upon well-defined inputs and outputs at the physical locations of interconnection between the systems. Conditions needed for interconnection at the component level must be known. Well-defined inputs and outputs usually imply that inputs and outputs are physical quantities that are known and measurable. Component parameters must match at locations at which the output of one component becomes the input of another.

## 5.4    Systems Are Dynamic

A characteristic of all types of systems is they are dynamic. Systems exist in a condition of continuous activity and change. Inputs to the system are transformed into outputs. Components internal to the system interact. Conditions within the system may change over time. Component interactions may be cyclic in nature or otherwise time dependent. Interactions with entities external to the system may change over time. Continuous activity characterizes systems.

Like all systems, technological systems are dynamic by their very nature. Some type of input crosses the system boundary. The dynamic interactions of the system components convert the inputs into the desired outputs. The time scales characterizing system action vary considerably across different types of technological systems.

Technological systems being dynamic does not necessarily require that system components are physically moving, or some type of visibly perceptible actions take place. The transfer of information through a computer network is a dynamic but invisible process. Alternatively, a non-moving wind turbine is not converting any energy of motion into electrical energy.

It appears reasonable to state that a technological system that is not in some way dynamic is non-functional or broken. A component failure or lack of a required input causes some interactions to cease. Absent, a critical subfunction or input, dynamic system activity ceases.

### 5.4.1  Time Varying Output

One aspect of the dynamic nature of technological systems is the status of the system and operating conditions can change over time.

Technological systems transform inputs into outputs often in a continuous process. These systems can be varying in time. The status and conditions at different locations in the system can change over time due to system operation and interaction with the environment. Component behavior can vary with time, such as degraded performance or wearing down, that can cause changes in system behavior and output.

A photovoltaic solar farm converts energy from sunlight into electrical energy. A typical utility-scale solar farm with a capacity of several megawatts, is shown in Fig. 5.11. This system has no perceptible motion when in operation. Even when the system is operating in its normal stable mode of operation, the amount of electrical energy output is not constant. The electrical energy output is dependent upon the solar energy input which varies continuously due to time of day, weather, and seasonal varying position of the sun in the sky.

On a clear day the output initially increases reaching a maximum at solar noon and then decreasing to sunset. If intermittent cloud cover occurs that will affect output as well. Figure 5.12 shows typical system output electric power on an entirely sunny day and also a day with intermittent cloudiness. In most locations seasonal variations will result in gradual daily changes in the length of the day as well. These combined effects resulting

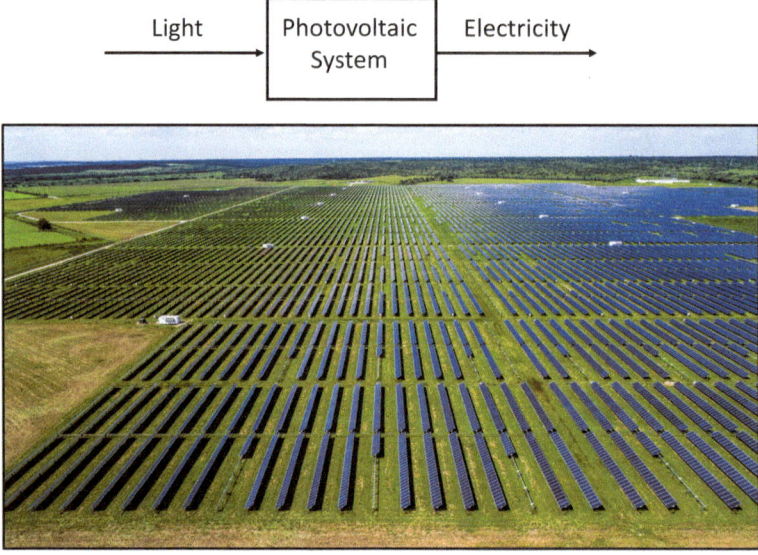

**Fig. 5.11**  Utility-scale solar farm (iStock.com/RoschetzkyIstockPhoto)

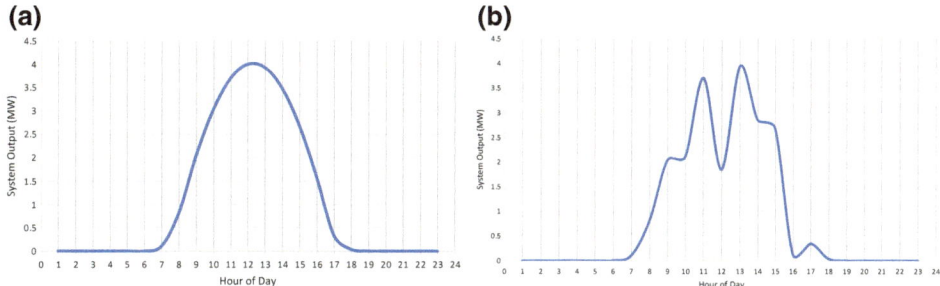

**Fig. 5.12** Representative solar farm output over time on: **a** clear day; **b** partly cloudy day (diagrams by author)

in dynamic input amounts of solar energy result in corresponding dynamic output of electrical energy.

Changes in component functioning can influence the dynamic behavior of technological systems. This can lead to time varying system output even if inputs and interactions with the environment are constant.

In the case of the photovoltaic solar farm the efficiency of the photovoltaic cells decreases as they heat up. System output decreases because the hot photovoltaic cells produce less electrical energy output for the same input.

Some solar farm installations are located in environments that result in a gradual accumulation of dust and dirt on the photovoltaic surfaces. The dust and dirt coating reduces the amount of sunlight hitting the photovoltaic cells. The overall system output decreases as the cells become more dust coated.

## 5.4.2 Multiple Modes of Operation

Many technological systems, especially major consumer products, have multiple modes of operation. In different operating modes different combinations of components can be active. Interactions between components and between the system and the environment can be considerably different in different modes of operation. Different operating modes is one aspect of the dynamic behavior and time variability of these systems.

A smartphone is an example of a consumer technological system with different modes of operation. The same physical object has multiple operating modes including sending text messages, displaying videos, taking photographs, serving as a web browser, acting as a flashlight, and running special purpose applications. These different operating modes will utilize a subset of all of the components within the system.

### 5.4.3   Different Phases or Stages of Operation

Technological systems can carry out different phases or stages of operation in the process of completing a single function. In the different phases, the components involved and the component interactions can be different. In the course of completing a single overall function, internal components and interactions are variable. Phases or internal stages of operation illustrate another dynamic aspect of technological systems.

A common domestic washing machine is a representative example of a system with different operating phases while completing a single overall function. This system goes through several different phases of operation in the process of carrying out one overall function of removing dirt from clothes. Different combinations of internal components participate in these different phases.

A representative domestic washing machine is shown in Fig. 5.13. From the consumer perspective the washing machine has a single overall function, that is to wash clothes. The end user expects one function. To complete the one function, multiple different internal modes occur in the washing machine system.

Initially the consumer loads clothes into the wash tub, adds detergent, sets the controls, and initiates the operation. In the first phase the washer is filling. This shown in Fig. 5.14. The inlet valves are open to allow water into the machine. The water level sensor provides a signal dependent on water level to the controller. While the tub is filling with water, no other components are active. The filling of the tub illustrates a dynamic feature common

**Fig. 5.13** Representative domestic washing machine (iSt ock.com/Bet-Noire)

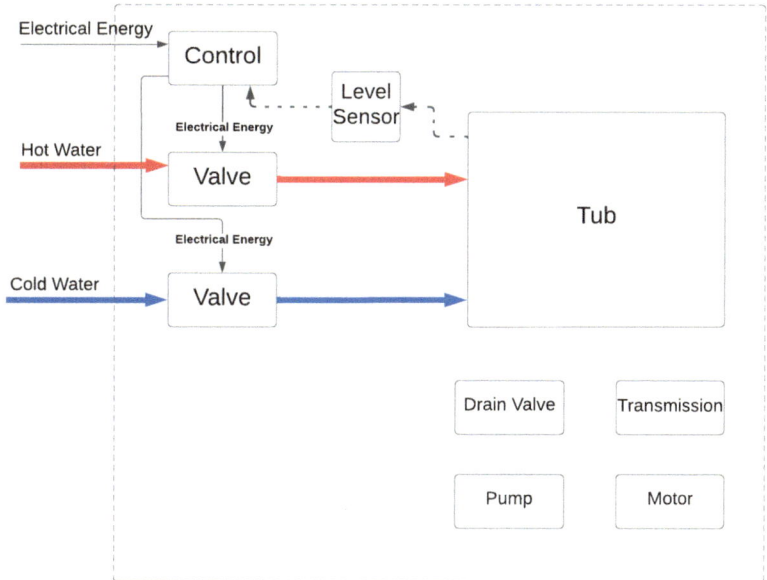

**Fig. 5.14**  Washing machine system in fill mode of operation (diagram by author)

in some systems, some type of time-dependent accumulation occurs within a component that influences system operation.

In the wash phase, the motor and transmission are functioning, the inlet valves are not involved. Later in the drain phase, the drain pump and motor are the major active components. These modes of system operation are illustrated in Fig. 5.15.

The Honda Prius hybrid vehicle shown in Fig. 5.16 illustrates significantly different component interactions in different operating modes. The driver is largely unaware of the different interactions taking place as all have the same result of moving the vehicle forward. The purpose of these different modes is to maximize the fuel efficiency of the vehicle.

When starting out the Prius operates in starting mode. The vehicle uses energy stored in a battery to provide power to an electric motor that drives the wheels. The internal combusting engine is not operating. Figure 5.17 shows the drive train component interactions during this operating mode.

Above about 15 mi/h (24 kph) the Prius power train switches to cruising mode. This is shown in Fig. 5.18. The internal combustion engine is ON. A combination of the engine and the electric motor provides power to wheels in a way that maximizes engine efficiency. The internal combustion engine is operated at its most efficient speed. Some engine power is used to run the generator which provides power to the electric motor. The goal is to maximize overall efficiency by a combination of the combustion engine and the electric motor.

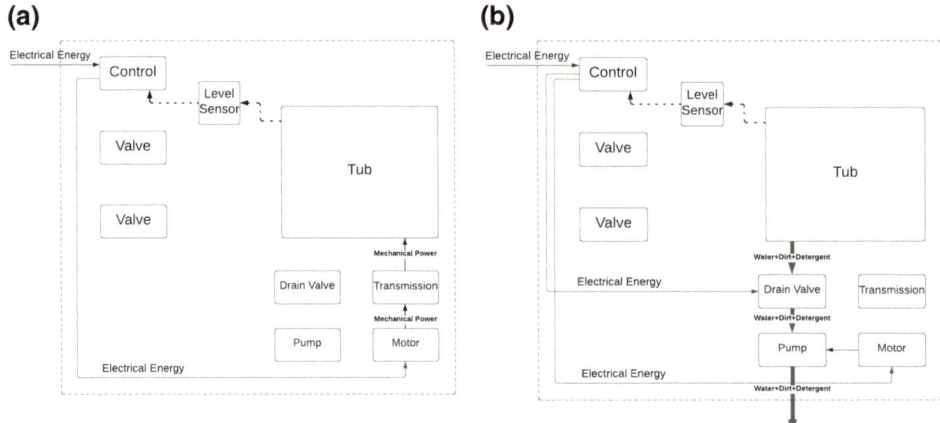

**Fig. 5.15** Washing machine system in wash and drain modes of operation: **a** wash mode; **b** drain mode (diagrams by author)

**Fig. 5.16** Toyota Prius hybrid vehicle [cropped] (iStock.com/contrastaddict)

### 5.4.4   Dynamic Feedback and Control

One aspect of the dynamic behavior of all types of systems is the concept of feedback and feedback control. Feedback describes the situation in which flow loops exist within the system that influence system dynamic behavior. Some systems contain loops internal to the system in which aspects of the conditions at a "downstream" location are sent back "upstream" or toward an input location in the system to influence "upstream" activity. These flow loops generally lead to dynamic, time-varying behavior.

Feedback loops or cycles are seen in many types of systems including ecological, geological, weather, and economic. For example, in the water cycle evaporation of water from surface sources evaporates which then condenses to form clouds. This leads to precipitation feeding back into surface reservoirs. The relationship between an animal population and food supply is another example of a feedback loop in a system. The size of food supply affects the animal population and the food consumption by the animal population is

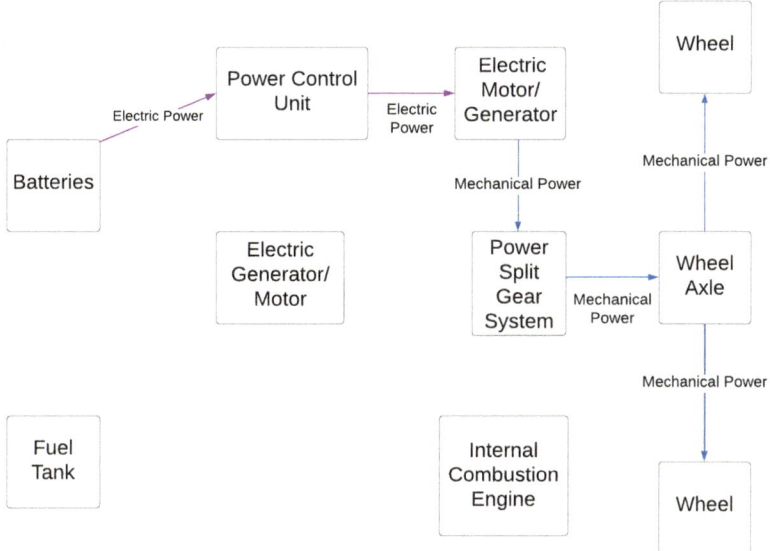

**Fig. 5.17** Toyota Prius starting mode drive train component interactions (diagram by author)

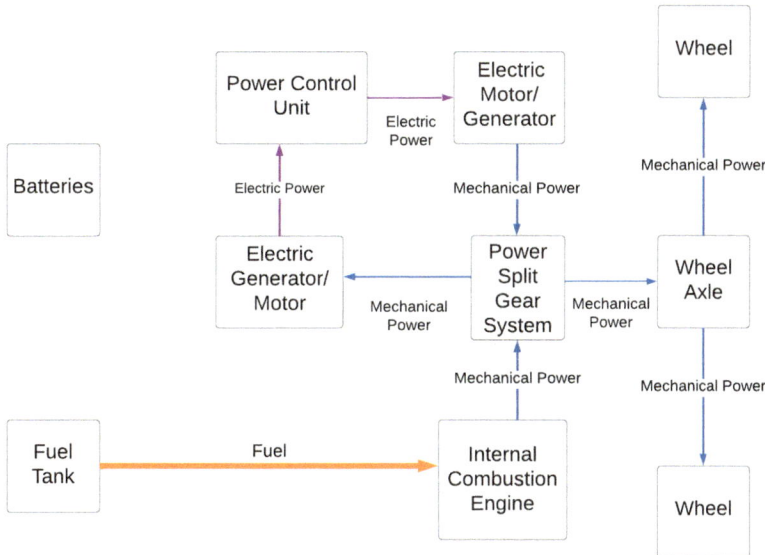

**Fig. 5.18** Toyota Prius cruising mode drive train component interactions (diagram by author)

one factor influencing food supply. A consumer products enterprise can utilize customer surveys as feedback information to adjust product characteristics and rates of production.

In many situations flows within a technological system are unidirectional starting from the inputs and moving to the outputs. However, technological systems can contain loops within the system in which material, energy, or information from a "downstream" location is sent back "upstream" or toward an input location in the system. Some feedback loops are used to provide automatic control of the system. Dynamic behavior of some systems can be attributed to the operation of these feedback loops.

Many technological systems include some type of automatic control of system operation. This relieves the user of some control functions. Automatic control often employs feedback loops within the system. System behavior if often maintained in a dynamic equilibrium within an acceptable range of operation.

The filling operation of an automatic washing machine shown in Fig. 5.14 above illustrated a basic example of a loop within the system. System operation is initiated, and the control system opens the inlet valves to allow water to flow into the tub. In the tub is a water level sensor. The sensor sends a signal back to the control system indicating the amount of water in the tub. Initially the level is low, and the control system leaves the inlet valves on. When the water level sensor signal indicates a full condition, the controller turns off the inlet valves stopping any further flow of water into the tub. In this way "downstream" information about water level loops back "upstream" to influence inlet valve operation.

A simplified version of a heating system demonstrates dynamic operation occurring from a feedback loop. Figure 5.19 is a diagram of one common type of heating system. The basic heating system has a furnace in which a combustion process takes place between air and fuel. This converts the chemical energy of fuel to thermal energy of hot exhaust leaving the furnace. The hot exhaust in turn heats a flow of room air moved through a heat exchanger by operation of a fan when the furnace is on. In the heat exchanger thermal energy is transferred from hot exhaust to the room air. The system utilizes feedback control using a temperature sensor located in the room. The sensor provides input to a control system that turns the furnace and air flow on and off.

Assume the thermostat is set for 70 °F (20 °C). The way the system works is when the temperature measured by the sensor falls to 2 °F below the set point or 68 °F (19 °C). The control system turns on the furnace and starts the flow of air from the room through the heat exchanger. Because there is some delay due to the time needed to heat up the heat exchanger, heat the airflow and cycle pre-existing cold air out of the vent, the room air temperature may drop further before starting to increase.

As warm air from the furnace enters the room, the temperature measured by the sensor increases. The rate of increase and the time needed to heat the room will depend on size of the room, the location of the sensor, and the rate of hot air flow from the furnace. When the sensor measures a temperature of 70 °F (20 °C), the control system turns off the furnace and may circulate the air for a brief time after the furnace is off to remove

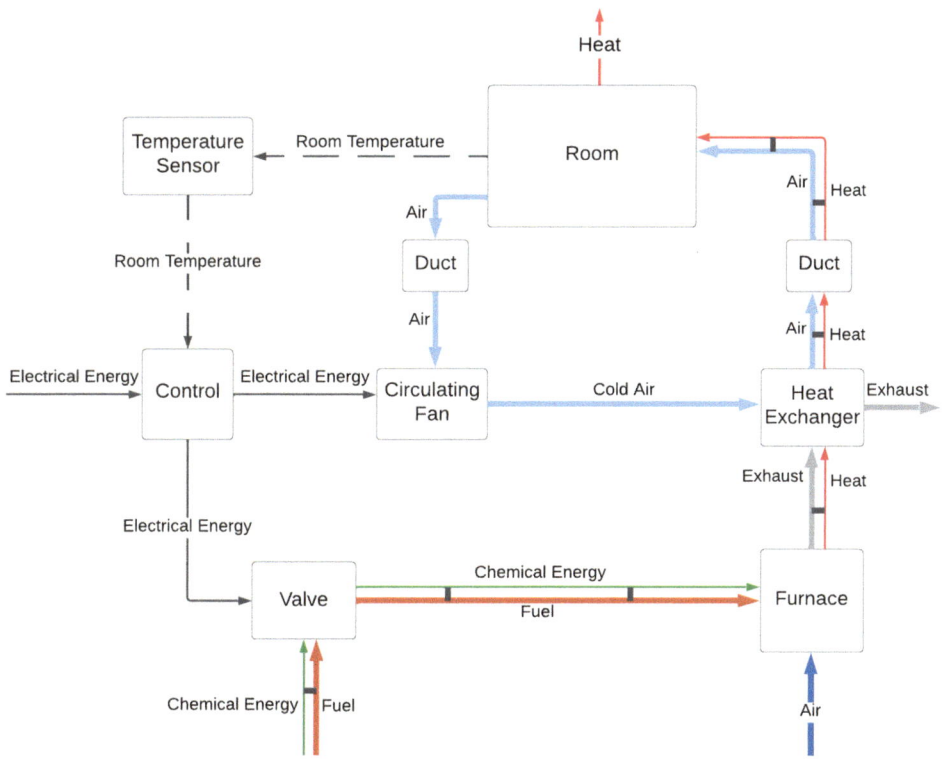

**Fig. 5.19**  Home heating system with feedback control interactions (diagram by author)

the heat remaining in the heat exchanger. The sensor may measure a temperature slightly above 70 °F (20 °C) after the air stops circulating.

Assuming it is still colder outside than inside, heat will flow from the room to the environment causing the temperature to decrease. When the temperature falls below 68 °F (19 °C) as measured by the sensor, the heating mode of operation will resume. Note that the colder it is outside the faster the rate of decrease in room temperature and the sooner the heating mode will restart.

The heating system illustrates dynamic time-varying behavior of a system. Control is accomplished by feeding back a sensor measurement to a control system. Time delays in the system response occur due to the time needed for the amount of energy to increase or decrease within a system component.

In addition to demonstrating feedback control, the examples of washing machine filling and the home heating system show how the dynamic behavior of systems can depend upon the rate of flows between components and the capacity of some system components. The time-dependent condition of the washing machine during the fill cycle depends on both the rate of water flow into the tub and the size of the tub itself in terms of water capacity.

In the heating system dynamic behavior depends on the size of the room being heated. A larger room with more air volume may take longer to raise the temperature. However, the dynamic behavior depends on the rate at which the furnace is transforming chemical energy into thermal energy and the rate of air flow through the heat exchanger. Additionally the rate of heat leaving the room such as being conducted through the building walls to the cold outside also influences the dynamic behavior of this system.

Technological system dynamic behavior is influenced by the rate of transfer of energy, material, or information between system components and the capacity of components that may be storing energy, material, or information.

## 5.5   Technological System Homeostasis

Homeostasis is term that is often applied to biological systems such as the human body. Homeostasis describes the ability and tendency of the system to maintain a relatively constant internal state in spite of changes in external conditions. Homeostasis refers to a system's ability to maintain a certain state of equilibrium. Biological systems achieve equilibrium through homeostatic control mechanisms.

The concept of homeostasis can be applied in an approximate way to technological systems to describe robustness of system behavior. Technological systems have a range of operating conditions and parameters within which they are able to operate as designed. Within certain limits, a system is able to produce the desired outputs with fluctuations or variations in inputs or component operation. This describes the concept of operating range or tolerance for disruptions while maintaining acceptable operation.

For technological systems the concept of operating range is typically considered as part of the system design process. Maintaining acceptable operation within defined limits can include both the intrinsic properties of components and the use of active control mechanisms to sustain operation.

A smartphone is specified as able to operate satisfactorily in temperatures between 0 and 35 °C (32–95 °F). If the phone experiences temperatures above 35 °C (95 °F) an internal control system shuts phone down to avoid damage.

Acceptable limits and ranges of operation are inherent in technological systems. There are many representative examples. The Golden Gate Bridge in San Francisco has a weight limit of 80,000 pounds for any truck using the bridge. A typical child's car seat, rear facing, can accommodate child weighing from 5 to 40 pounds. A GE General Electric Haliade 150–6 MW Wind turbine can operate in wind speeds from 3 to 25 m/s (7–56 mph).

## 5.6    Nonlinearity

Nonlinearity is a characteristic behavior of some systems in which a response output is not in simple proportion to a change in input. Under certain conditions, a small change in a system input or interaction produces a considerably magnified change in some status condition of the system. Behaviors such as reaching a "tipping point" or S-curve effects are nonlinear outcomes. Due to the disproportionate effect of small changes under specific conditions, nonlinearity can be a particularly critical feature of system behavior.

Some technological systems, like other types of systems, can have nonlinear behavior.

The common circuit breaker exemplifies nonlinear behavior. Figure 5.20 shows circuit breakers in an electrical distribution panel and an individual circuit breaker. The circuit breaker protects circuits from excessive current. A circuit, for example, may have a maximum safe limit of 15 A amperes. When currents between 0 and 15 amperes exist in the circuit, the circuit breaker does nothing other than allow the current to pass through. If a current reaches 15 amperes, the circuit breaker quickly activates to open the circuit and stop the current.

This is an example of nonlinear system behavior. Over the wide range from 0 to 15 amperes nothing happens, but a slight increase in current near 15 amperes creates a sudden

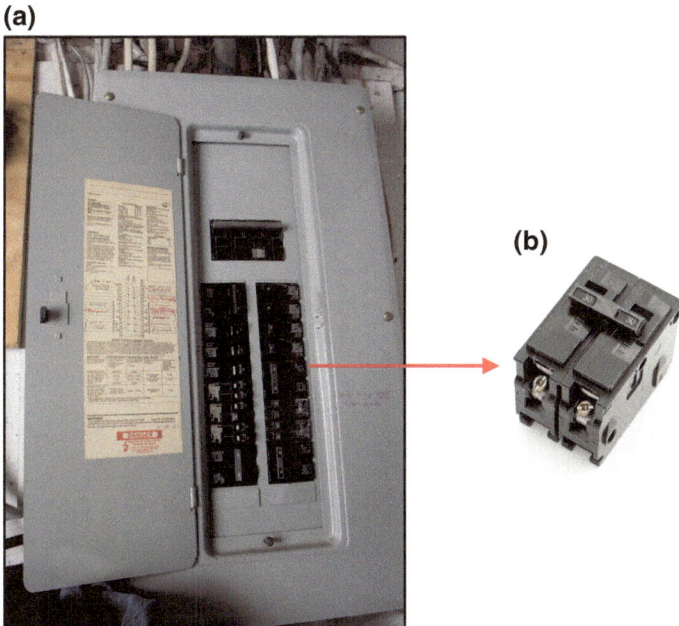

**Fig. 5.20**  Circuit breaker: **a** breaker box (iStock.com/KyleNelson); **b** circuit breaker (iStock.com/ DonNichols)

**(a)**

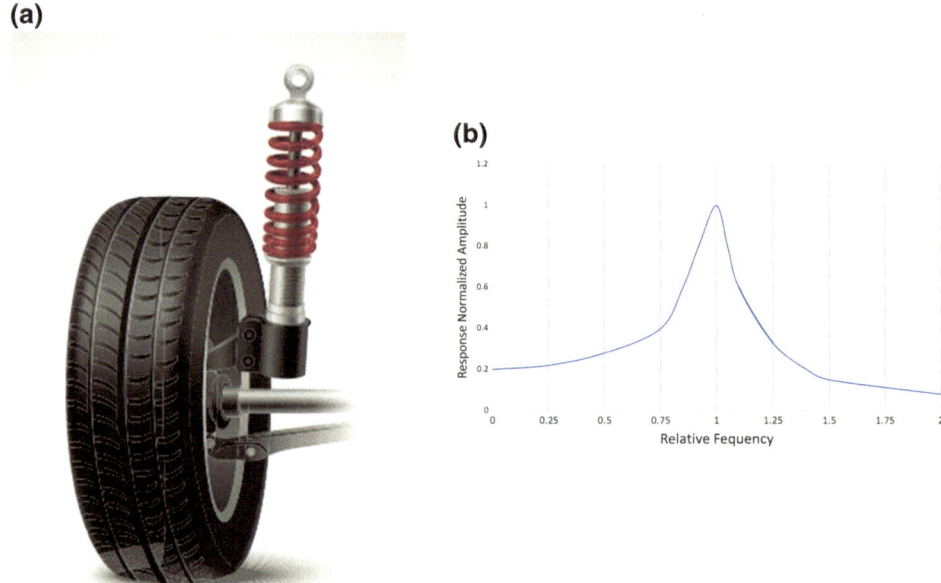

**Fig. 5.21**  **a** Automobile suspension system (iStock.com/Macrovector); **b** characteristic impulse response curve (graph by author)

and significant change in system operation. The abrupt nonlinear behavior of the circuit breaker is central to its function of protecting circuits from current above a particular expected amount.

Automobile suspension systems can potentially have nonlinear behavior. One function of the suspension system is to absorb bumps from the road by flexing. As shown in Fig. 5.21, in some circumstances the suspension response can be significantly more pronounced displaying a nonlinear characteristic. Design of suspension systems makes a deliberate effort to reduce the occurrence of this nonlinear system behavior.

## 5.7    Component Autonomy

System components are autonomous in the sense that the internal processes within components are independent of the other components. The actions taking place within a component are determined by the system requirements and the component inputs and outputs. As long as a system is functioning and the necessary component-to-component interactions are taking place, the components are "indifferent" regarding the processes within the other components.

System component internal independence is an aspect of the system properties of emergence and hierarchies. The system functions emerge from the interactions of the system elements. As long as the elements carry out subfunctions, accept inputs, and produce specific expected outputs, the overall system functions as expected.

The system property of component internal independence is demonstrated differently in various systems. In organizational and managerial systems, the phenomenon of "siloing" reflects the situation in which different system elements may be oblivious to activity going on in other elements. In a complex biological ecosystem, events affecting one element, such as a disease, only effect the other elements to the extent that interactions are changed.

In technological systems, the concept that components are internally independent leads to a diversity of physical phenomena being utilized in most systems of even moderate complexity. Relatedly, components with markedly different internal processes can be utilized in a technological system provided that they can exchange at least one common flow of materials, energy, or information.

## 5.7.1  Diversity of Underlying Effects

The complete operation of most technological systems includes a diverse range of components employing a wide assortment of physical principles not confined to any one area of science. A technological system is not restricted to only use phenomena from one specific area of science. Physical phenomena are classified in a particular logical order by science. Most technological systems employ phenomena from widely separated areas of science.

An internal combustion engine automobile illustrates the wide range of different types of physical effects that can be employed in a technological system. Consider just four representative components as shown in Fig. 5.22. These are: the starter motor, the brakes, a catalytic converter, and the tires. These components are essential to system operation but have little relation to each other in terms of the internal phenomenon utilized by the component.

The starter motor begins the engine crankshaft turning so the four-stroke cycle based on combustion of fuel and air can begin. The starter is an electric motor that creates rotation by application of magnetic forces. The starter is supplied with electric current from the battery. The starter motor is shown in Fig. 5.23.

The brakes slow and stop the vehicle using friction as illustrated in Fig. 5.24. Friction is caused by the intermolecular attraction between molecules on the surface of objects. Friction is created by the interaction of the brake pads with the brake rotor.

The catalytic converter reduces pollution from an internal combustion engine by causing chemical reactions in the exhaust. It is usually located along the underside of the vehicle as shown in Fig. 5.25. In the catalytic converter chemical reactions change pollutants of carbon monoxide, unburned hydrocarbons, and NOx into water and carbon dioxide.

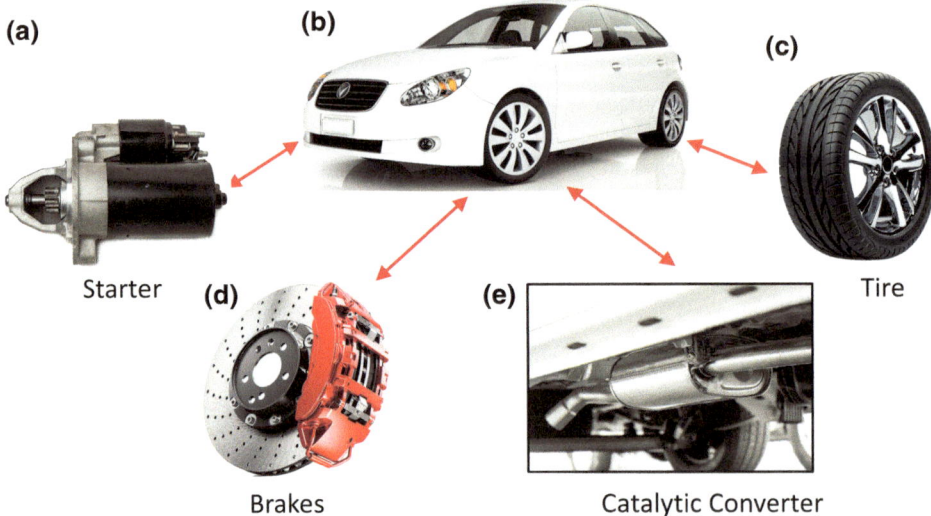

**Fig. 5.22** Sample automobile components: **a** starter (iStock.com/Aleksandr Kondratov); **b** car (iStock.com/Rawpixel); **c** car wheel (iStock.com/BlackJack3D); **d** car brakes (iStock.com/Grassetto); **e** catalytic converter (iStock.com/deepblue4you)

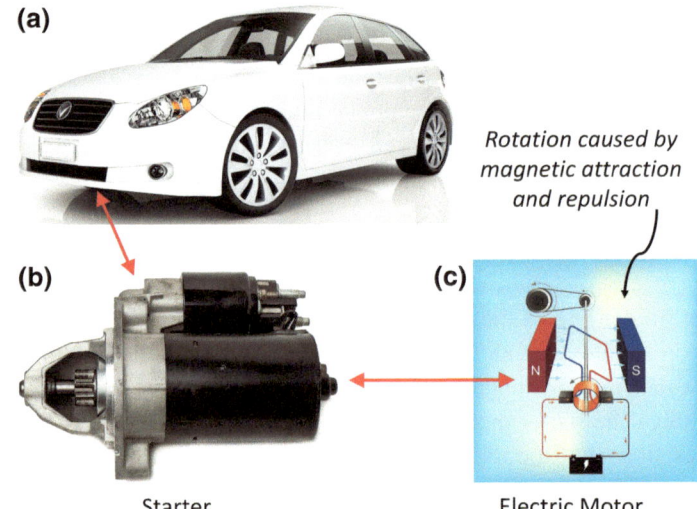

**Fig. 5.23** Starter based on magnetic forces: **a** car (iStock.com/Rawpixel); **b** car starter (iStock.com/Aleksandr Kondratov); **c** DC generator (iStock.com/Graphic_BKK1979)

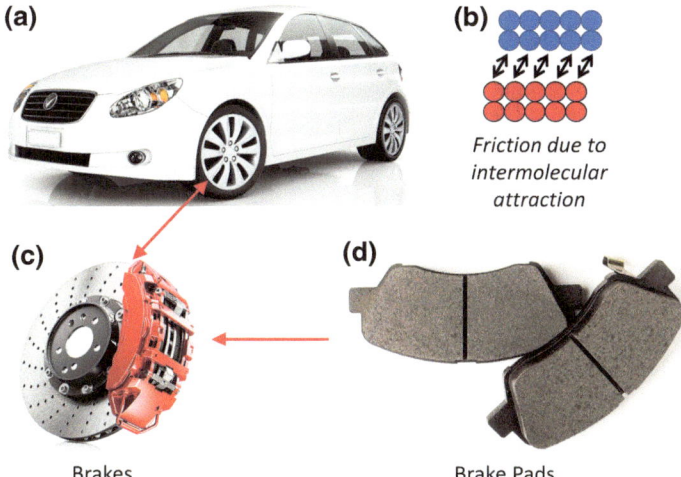

**Fig. 5.24**  Brakes using intermolecular attraction of friction: **a** car (iStock.com/Rawpixel); **b** friction illustration (illustration by J. Krupczak); **c** brakes (iStock.com/Grassetto); **d** brake pads (iStock.com/vav63)

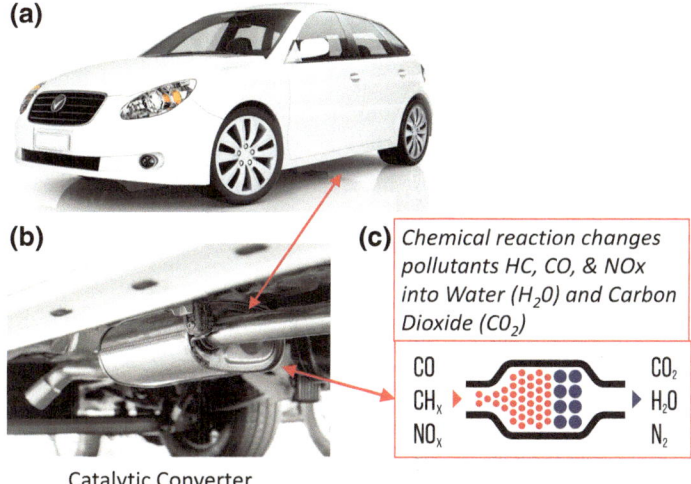

**Fig. 5.25**  Catalytic converter using chemical reaction to reduce pollutants: **a** car (iStock.com/Raw pixel); **b** catalytic converter (iStock.com/deepblue4you); **c** catalytic conversion diagram (iStock.com/Art_rich)

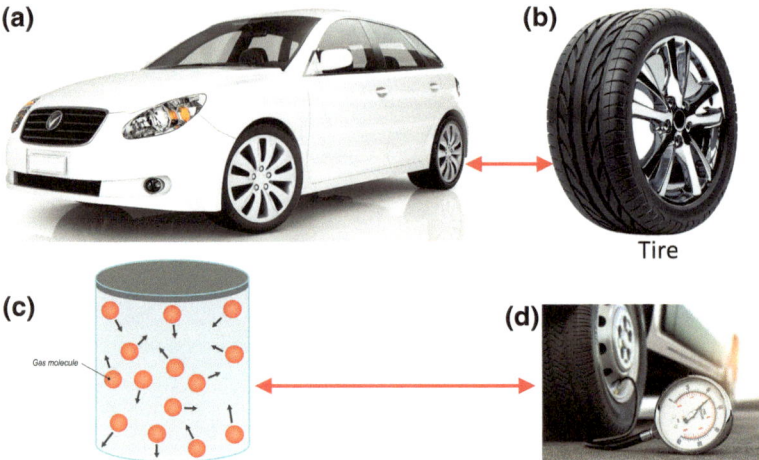

Tire pressure due to collisions of
air molecules with inside surface

**Fig. 5.26**  Tires support weight relying on collisions of air molecules: **a** car (iStock.com/Rawpixel); **b** tire (iStock.com/BlackJack3d); **c** Kinetic Theory of Gases (iStock.com/Honourr); **d** tire pressure gauge (iStock.com/Bet_Noire)

The ability of tires to support the weight of the car is due to the tire pressure. The pressure is determined by the rate at which molecules in the air inside the tire collide with the tire walls. Figure 5.26 illustrates the source of pressure as molecular collisions. The tires are inflated to a specified air pressure which is determined by the density of air molecules in the tire volume and the temperature.

These four components of the automobile demonstrate the range of physical effects that are applied to fulfill a specific function in this technological system.

Figures 5.22 through 5.26 illustrate the use of magnetic forces, intermolecular attraction, chemical reaction, and collisions of air molecules in the operation of the automobile. The automobile example conveys how no particular phenomenon has intrinsic value that demands or ensures that it be utilized in the automobile. Each phenomenon is employed by virtue of the function provided and the contribution to the overall system function of the automobile. The starter motor, brakes, tires, and catalytic converter contribute subfunctions to overall system operation. Other phenomena could be utilized if they were better able to accomplish the overall function and subfunctions that are provided by magnetic forces, attraction of surfaces, chemical reactions, or molecular collisions in these particular components. In a technological system the value of a phenomena is determined by the ability to provide a useful function while meeting relevant requirements of the system.

## 5.7.2  Same Flow Different Effects

Components of the system that utilize different physical principles may interact provided they share a common flow of the same material, energy, or information. Although technological systems involve components that are based on different scientific principles, these components, based on diverse phenomena, can and do interact with one another by the exchange of materials, energy, and information in achieving overall system function.

In a technological system two components can interact so long as they are exchanging the same flow of materials, energy, or information. Two components interact when the output of one component serves as the input to another component. What is going on inside one component has no bearing on what is taking place inside the other component. The only issue of significance is that one component's output can be used by the other component as an input. An alternative viewpoint is if two components interact they must share at least one common flow of material, energy, or information.

The refrigerator illustrates how components using markedly different underlying effects interact via the flow of material, energy, or information between these components. The central effect put to use in keeping food cool is the principle of evaporation. To evaporate and change state from a liquid to a gas, energy must be absorbed by the liquid. In the refrigerator, the evaporating liquid is the refrigerant that flows within tubing in the refrigerator walls. By losing energy to the evaporating refrigerant liquid, the temperature of the refrigerator interior is reduced. This takes place in the evaporator component that is a coil of serpentine shape through which the refrigerant flows to absorb heat and evaporate.

After the refrigerant evaporates it must be increased in pressure as part of the process of returning it to the liquid state. This is accomplished in the compressor. The compressor uses a pushing force from a fast-moving piston to squeeze the gas molecules from the evaporator closer together. This increases the pressure and also pushes the gas through the system to the next stage of the process.

The evaporator and the compressor utilize different principles to perform their respective functions; however, they share a common interaction. The output of one is the input of the other. The refrigerant vapor leaving the evaporator serves as the input to the compressor. The evaporator and compressor, while different in principle and function, both interact with the flow of the refrigerant material.

Although the main components of the refrigerator are based on different underlying phenomena, these different phenomena interact via the exchange of materials, energy, or information. The compressor uses an electric motor based on the principle of magnetism and the evaporator uses the principle of evaporation. These two principles interact by means of the flow of a material, the refrigerant liquid, which flows between the evaporator and the compressor.

The refrigerator door illustrates a different aspect of component independence. The door of the refrigerator must be easy to open, yet when closed it must seal tightly so that warm air and moisture are not drawn into the refrigerator. Most modern refrigerators have a magnetic seal around the perimeter of the door. The door contains the magnet and, when closed, the magnets are attracted to the steel of the door. This makes use of the phenomenon of magnetism. Some materials, in this case steel, experience a force when placed in a magnetic field. This force keeps the refrigerator door closed. The magnetic door seal is identified in Fig. 5.27.

The function of holding the door closed does not have to be accomplished using magnetism. Earlier refrigerators used a latch mechanism. The latch seal was based on a mechanical clasp. Magnetism replaced the latch because it is easier to open, is less prone to gaps in the seal, is quieter, and has no latch mechanism to break. The function remained unchanged, that is holding the refrigerator door closed. The means to achieve this function is now accomplished by applying a different principle. Figure 5.28 shows the type of refrigerator door latches that were once common.

**Fig. 5.27** Magnetic refrigerator door seal (iStock.com/Alexey Rotanov)

**Fig. 5.28** Vintage refrigerator door latch (iStock.com/Alexey Rotanov)

As refrigerator door closures changed from mechanical latches to magnets the evaporator and compressor remained unchanged. The evaporator and compressor do not interact directly with the door. There is some indirect interaction in the sense that the door should generally be well-sealed to reduce heat flow into the refrigerator and not exceed the cooling capacity of the evaporator. Components that do not share an interaction are independent of each other. The evaporator and the compressor "don't care" how the door is held closed.

The components of a technological system may be based on different underlying effects or phenomena, but to contribute to the overall function of the system the components must interact in some manner. Components must interact with other components to contribute to the operation of the system. If two components interact they must share at least one common flow of material, energy, or information, even though the transformation taking place in those components can be due to substantially different underlying effects.

## 5.8    System Failure Analysis

The system property of emergence and the hierarchical nature of systems suggest an approach to failure analysis frequently used in many types of systems. Failure or non-operation is typically caused by elements not functioning as required or lack of a required input to a component. However, identifying non-functioning elements is not necessarily straight-forward in various types of systems that might have a large number of interacting elements, multiple modes of operation involving different interactions in each mode, and complex arrays of inputs and outputs during different aspects of system operation.

The hierarchical nature of systems in general results in most systems existing as a "system of systems." Efficient failure analysis first identifies the subsystem most likely to be responsible for failure. Analysis can then proceed to the component-level diagnosis within the subsystem. This approach provides information about the conditions and status within the system which may be a challenge to acquire under failure conditions.

The human body is an example of a "system of systems." The diagnosis and treatment of disease and illness is an example of how system hierarchies naturally foster an approach to failure analysis. Diagnosis first identifies the major system in the body's "system of systems." These include the musculoskeletal system, the respiratory system, the circulatory system, the digestive system, the nervous system, the endocrine system, and the reproductive system. Within the subsystem, analysis then proceeds to identification of specific affected organs and tissues. In medical analysis of the human body system, significant effort is often needed to obtain information to fully identify the nature of the nonfunctional condition. This would include, for example, diagnostic tests such as blood tests, kidney function test, EKGs, X-rays, and MRIs.

As is the case with the human body as a system, analysis of technological system failure frequently exploits the hierarchical nature of technological systems. The functionality

**Fig. 5.29** Flat tire example of a simple component failure disrupting entire system function (iStock.com/Jamesb owyer)

of a technological system emerges from the interactions of the constituent components. The components working together can provide a utility beyond the mere "sum of the parts." A converse to this is failure of any one component can exert what might seem as a disproportionate influence in disrupting operation of the entire system. The flat tire illustrated in Fig. 5.29 is an example. A relatively simple failure in a relatively uncomplicated component disables the entire system.

Failure analysis utilizing the hierarchical system-of-systems nature of technological systems as a general approach is demonstrated even in consumers' routine interactions with technology. As an example, the electric power system is one of the most spatially expansive technological systems spanning an entire country. The hierarchical organization begins with hundreds of generating stations and proceeds through multiple levels interconnected subsystems eventually terminating in millions of individual wall outlets.

Familiar failure analysis illustrates both some of the levels of the national electrical system hierarchy and application of these levels in problem solving. Figure 5.30 illustrates this example. When failure is observed in one single component, the problem is suspected to be at the lowest level of the hierarchy. For example, if the light is not working, suspect the bulb and replace it. If multiple electrical appliances in the same room simultaneously fail, then the next higher level of system hierarchy is suspected, the circuit providing power to all outlets in the room. In this case, a circuit breaker failure is a possible problem. If all electrically powered systems in the entire structure simultaneously stop, and perhaps also some nearby residences are without electric power as well, then problem analysis proceeds to the hierarchical level associated with neighborhood distribution.

Failure analysis in all types of systems depends upon information characterizing which interactions within the system are inoperative and, relatedly, which components are not providing expected subfunctions. As noted in the case of the biological system that is the human body, a considerable aspect of diagnosis and treatment is often obtaining diagnostic information through tests and measurements about conditions within the body system. Those conducting failure analysis in technological systems face the same issue of obtaining information needed to identify and remedy non-working components. In some technological systems, problem analysis is complicated by a complex system which provides few visual indicators of system status.

(a)                          (b)                          (c)

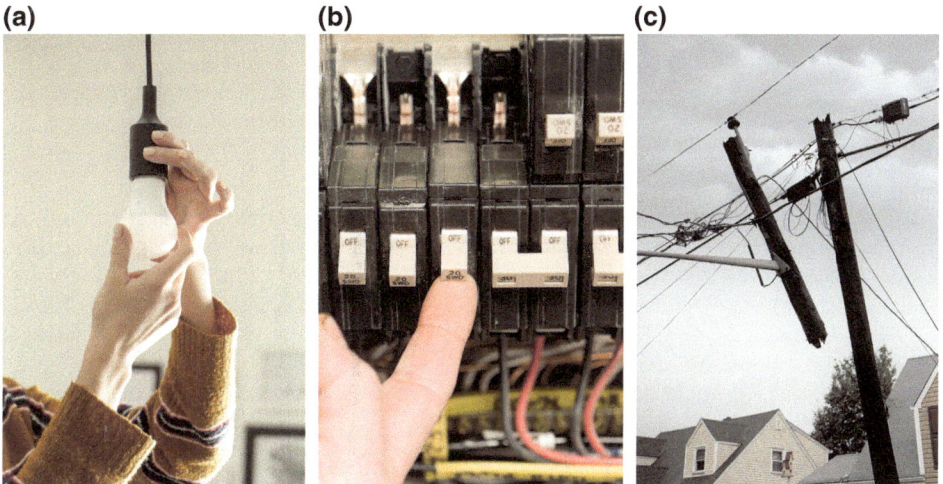

**Fig. 5.30** Failure analysis at different hierarchical levels of the electric power distribution system: **a** changing lightbulb (iStock.com/triocean); **b** circuit breaker (iStock.com/BanksPhotos); **c** mangled utility pole (iStock.com/Bluberries)

To assist failure analysis, some technological systems include components specifically to facilitate determining causes for failure in non-functioning systems or components. For example, most automobiles include an "on-board diagnostics" or OBD system to assist in failure analysis. In operation, many automobile systems include sensors that might be used to control some aspects of operation. These might include engine or motor temperature, coolant temperature, oxygen levels, engine or motor speed, and battery condition. The OBD system records some sensor information and also records off-normal occurrences in the form of "trouble codes." This information can then be "read back" by automotive technicians to diagnose faults other than drivability issues. A OBD Scanning device can

**Fig. 5.31** Automobile On-board Diagnostic (OBD) scanner reading diagnostic trouble codes (iStock.com/Bir dlkportfolio)

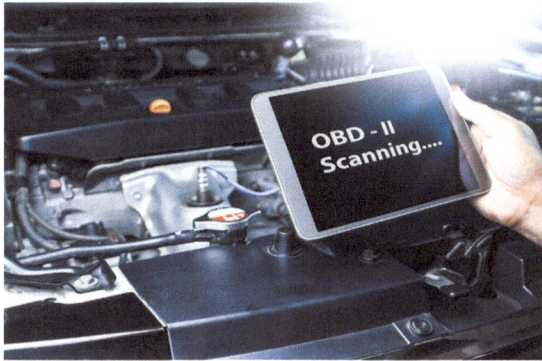

be connected to the car to retrieve the diagnostic trouble codes and recorded information as shown in Fig. 5.31.

## Bibliography

Adams, Kevin MacG., Patrick T. Hester, Joseph M. Bradley, Thomas J. Meyers, and Charles B. Keating. "Systems Theory as the Foundation for Understanding Systems." *Systems Engineering* 17, no. 1 (2014): 112–23. https://doi.org/10.1002/sys.21255.

Adams, Kevin MacG. "Systems Theory: A Formal Construct for Understanding Systems." *International Journal of System of Systems Engineering* 3, no. 3–4 (January 2012): 209–24. https://doi.org/10.1504/IJSSE.2012.052684

Arnold R.D., and J.P. Wade, "A Complete Set of Systems Thinking Skills," *INCOSE International Symposium,* vol. 27, no. 1, pp. 1355–70, Jul. 2017. https://doi.org/10.1002/j.2334-5837.2017.00433.x

Boulding, K., "General Systems Theory." *Management Science*, vol. 2, no. 3, pp.197-208, Apr. 1956.

Checkland, Peter. *Systems Thinking, Systems Practice.* John Wiley, 1999.

Forrester, J.W. *Principles of Systems*, Waltham, MA: Pegasus Communications, 1968.

Halderman, James. *Automotive Technology: Principles, Diagnosis, and Service.* Pearson/Prentice Hall, 2020.

Meadows, Donella H. *Thinking in Systems: A Primer.* Chelsea Green Publishing, 2008.

*MITRE Systems Engineering Guide.* The MITRE Corporation, Bedford, MA. 2014. https://www.mitre.org/publications/technical-papers/the-mitre-systems-engineering-guide.

NASA Project Gemini Familiarization Manual, Rendezvous and Docking Configuration, Orbital Attitude Maneuvering System, Control Number C-160063, July 1966.

National Renewable Energy Laboratory, "Solar Energy Basics." Accessed August 1, 2023. https://www.nrel.gov/research/re-solar.html.

Senge P., *The Fifth Discipline: The Art and Practice of the Learning Organization*, 1st ed. New York: Doubleday/Currency, 1990.

Simon, Herbert A. *The Sciences of the Artificial, Third Edition.* MIT Press, 1996.

SpaceX, Update: In-Flight Abort Static Fire Test Anomaly Investigation https://web.archive.org/web/20190719221112/https://www.spacex.com/news/2019/07/15/update-flight-abort-static-fire-anomaly-investigation.

Stave K., and M. Hopper, "What Constitutes Systems Thinking: A Proposed Taxonomy." *25th International Conference of the System Dynamics Society*, Jul. 1, 2007. https://digitalscholarship.unlv.edu/sea_fac_articles/201.

Stoecker, Wilbert F., and Jerold W. Jones. *Refrigeration and Air Conditioning.* McGraw-Hill, 1986.

Sweeney L.B., and D. Meadows, *The Systems Thinking Playbook.* White River Junction, VT: Chelsea Green, 2010.

*Systems Engineering Fundamentals.* Department of Defense, Systems Management College, Defense Acquisition University Press, Jan.2001.

Weinberg G.M., *An Introduction to General Systems Thinking.* New York: Dorset House Publishing, 2011.

Von Bertalanffy, L. "An Outline of General System Theory." *The British Journal for the Philosophy of Science,* vol. 1, no. 2, pp. 134–65, Aug. 1, 1950. https://doi.org/10.1093/bjps/I.2.134.

# Component Parameterization and Transfer

**6**

## 6.1    Chapter Overview

- A characteristic of the creation of modern technological systems is the use of the same component to carry out the same function in different technological systems.
- Component functions can transfer across systems. The need for a particular subfunction appears in a variety of technological systems.
- Families of components exist offering features suited for diverse situations while providing the same primary component function. The central physical principle or phenomenon at work in the component remains the same.
- For inclusion in a system, a component must provide desired functions while having form properties consistent with system requirements. Aspects of the form of existing components can be modified or altered to suit the requirements of different systems more appropriately.
- Established components typically become available with standardized features and properties such as standard physical dimensions or levels of functional performance.
- Standardization of components facilitates interchangeability as well as adoption of components into new or different technological systems.
- Component manufacturers enable component adoption by providing information about available components and their properties relevant to incorporation into systems.
- Component parameters of frequent interest include performance characteristics, operating ranges, interconnection parameters, and physical dimensions.
- Manufacturers' component information facilitates achieving function and form requirements for component incorporation into a system.

J. Krupczak, Jr., *Understanding Technological Systems*, Synthesis Lectures on Engineering, Science, and Technology, https://doi.org/10.1007/978-3-031-45441-7_6

## 6.2    Component Transfer Across Systems

The same component can be used to provide this subfunction in different systems. Different systems can have a need for the same subfunction within the overall system. In this way component functions can transfer across systems.

The idea of components providing transportable functionality is intuitive and widely recognized. A well-known saying is "do not reinvent the wheel." The message in this advice is components may already exist to carry out a necessary subfunction.

Figure 6.1 illustrates the principle of component transfer across different systems. The same type of caster wheel is used on the hospital bed and the food service cart. These two systems have different overall applications. One supports the weight of a patient in a hospital, the other supports trays of prepared food. Each of these different systems has the need for a component to provide the function of supporting a load while reducing friction.

The electric motor provides another example of component transfer across systems. The electric motor converts electrical energy into energy of motion. This function finds use in diverse applications. For example, an electric motor is used to rotate the shaft inside an electric drill. This is illustrated in Fig. 6.2. A similar motor is used in a hair dryer. In the drill the motor rotates the shaft, in the hair dryer the motor rotates the blades of the fan. The same electric motor provides the same function as a component used in distinctly different systems or applications.

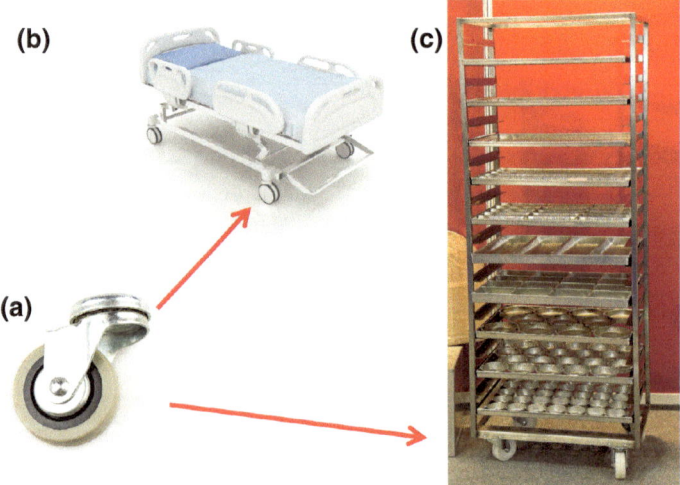

**Fig. 6.1** Caster wheel used in different applications: **a** caster wheel (iStock.com/kritsada pananchai); **b** hospital bed (iStock.com/rzelich); **c** food service rack (iStock.com/Baloncici)

**(a)**                                          **(b)**

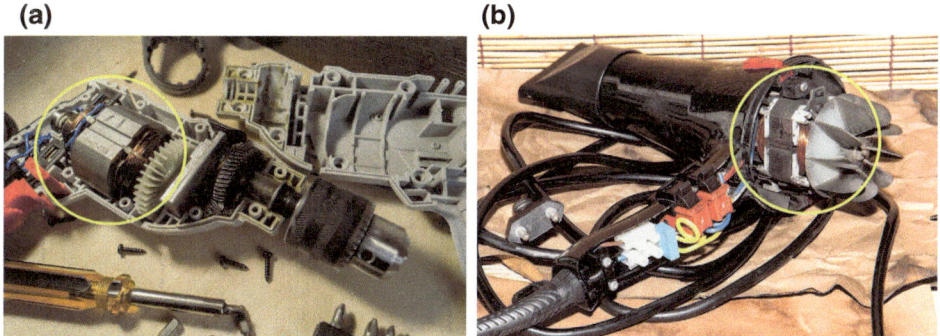

**Fig. 6.2** Electric motor used in different applications: **a** electric drill (iStock.com/Ivan Reshet-nikov); **b** hair dryer (iStock.com/theeraphon)

## 6.3    Subassemblies, Modules and Complex Component Transfer

Components need not be simple elements such as wheels or electric motors. The use of more sophisticated and complex components also transfers across systems.

Technological system components span a spectrum from simple to elaborate. The key characteristic of any component is that it provides a well-defined function and the ability to be integrated into different systems. That function may be utilized in a diversity of technological systems.

Complex components are sometimes called subassemblies, subsystems or modules. The component is still serving to carry out or contribute to the transformation of a portion of the system inputs into desired outputs. However, subassemblies are components that are typically more elaborate or multifaceted. More factors may need to be considered in utilizing the component in the system.

Like all components, subassemblies are recognized by their role as an element working in concert with other elements to produce the overall system function.

*Example: Touchscreen Display*
An example of a complex component or subassembly is a touch screen display as shown in Fig. 6.3. This type of display is used to convey visual information to the user of a device such as a smartphone and also receive input from the user by touching the screen.

The front of the display looks not very much different from what is seen by the user. The lower part of this component reveals the complexity of this subassembly. This display screen can be found in different types of smartphones and interactive devices.

**Fig. 6.3** LCD touch screen
display (photo by A.
Krupczak)

## 6.4    Component Integration Requirements

For inclusion in a system, a component must not only provide desired functions but
also interact compatibly with other components and have form properties consistent with
requirements of the system. Components are included in a system to carry out transformations of input to outputs and the scale and functional characteristics of this transformation
must be sufficient to achieve the purposes of the system. Components must also have form
features such as size and dimension to compatibly interconnect with other components.
In addition, requirements may exist for the form characteristics of the entire system that
must be respected by each system component.

Figure 6.4 illustrates the condition that a component must not only provide specific functional capabilities but must also successfully integrate with other requirements
imposed on the system. In this case, the component is a wheel. The basic function of the
wheel is appropriate for this particular system, i.e., the car. However, some characteristics
of the component are not compatible with the system requirements. First the component
appears barely sufficient to provide the main function of transferring the weight of the car
to the ground.

The component appears to lack form characteristics that are needed to meet some
requirements of the system. The wheel must support the vehicle load while rotating at
a high speed. The wheel should be able to sustain impacts from an uneven road and
main structural integrity. There are expectations for the useful operating life span of the
wheel that this component does not seem likely to meet. The need to interconnect with
other components such as the brakes is not accomplished. This example shows how a

**Fig. 6.4** Car with
non-matching wheel (This
image was created with the
assistance of DALL-E-2 by A.
Krupczak)

component integrated into a system must not only provide a specific needed function but
also have form properties addressing all requirements of the system.

## 6.5    Component Parameterization

A component can be modified or altered to suit the conditions of a particular system
more appropriately. Families of components exist which offer features suited for diverse
situations but provide the same primary function. The physical principle or phenomenon
at work in the component family remains the same for each individual component.

A basic component design can be adjusted to change the scale of the component
function. Aspects of the form features of a component might be modified to better suit
specific system requirements. Components are parameterized around a range of variables
to facilitate utilization in new systems.

### 6.5.1    Parameterized Functional Characteristics

Components in technological systems receive inputs from other components or the environment external to the system and transform these inputs in some way that advances
the overall system function. Components receive a certain amount of input and transform
this into some quantity of output. Different systems or different locations within a system may require different amounts of input and output. One common type of component
parameterization is to vary the magnitude of the input to output transformation.

**Fig. 6.5** Electric heater components used in electric water heaters: **a** 700 W heating element (cropped) (iStock.com/Serhii Ivashchuk); **b** 2000 W heating element (cropped) (iStock.com/Serhii Ivashchuk)

**(a)**             **(b)**

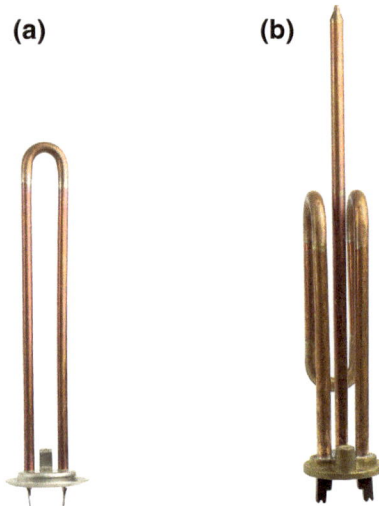

### 6.5.1.1   Component Variation Example: Electric Heater

An electric heater converts electrical energy into heat. The electrical energy input is transformed into heat output. The magnitude of this transformation can be quantified as a rate of energy conversion measured in Watts.

Figure 6.5 shows electric heater components used in electric water heaters. One is rated at 700 W. This component will produce heat from electrical energy at a rate of 700 W. Other applications may require a higher rate of heat production. The same basic component is available in a larger rating of 2000 W. As can be seen in the figure, the essential design of the component is similar in the smaller and larger wattage rating.

### 6.5.1.2   Component Variation Example: Electric Switch

Switches are used to actuate electric current in technological systems. Figure 6.6 shows one common type of electrical switch called a toggle switch. Toggle switches have a lever that the user moves from the "off" to "on" position to actuate the current. The switches are parameterized in terms of the maximum flow of electric current through the switch in the "on" condition. The first switch is rated at 5 amperes maximum current. The second switch can support a current of up to 10 amperes, while the third can have input to output current of up to 15 amperes.

### 6.5.1.3   Component Variation Example: Centrifugal Pump

The centrifugal pump provides another example of parameterized component magnitude in a moderately complex component. The centrifugal pump is used for transporting liquids such as water through piping systems. Applications include chemical processing, irrigation, and cooling systems. Centrifugal pumps transfer mechanical energy to pressure energy in the fluid using a rotating bladed impeller. The rotation is often provided

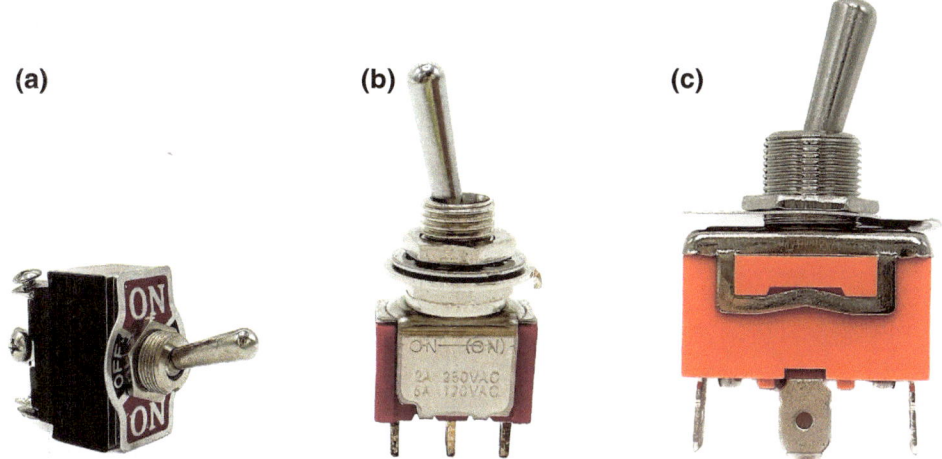

**Fig. 6.6** Electrical switches: **a** 5 amperes toggle switch; **b** 10 amperes toggle switch; **c** 15 amperes toggle switch (photo by A. Krupczak)

by an electric motor. The centrifugal pump can be viewed as converting electrical energy input into pressure energy of the output flow.

Figure 6.7 shows centrifugal pumps. Each pump provides different rates of flow through the pump with a specified electrical energy input. Key parameters characterizing the input to output transformation are rate of flow from input to output, output pressure, and electrical power input to the pump electric motor.

The first pump requires 2.24 kW of electrical energy input to provide 34 cubic meters per hour at a pressure of 0.124 MPa (150 gallons per minute output at pressure of 18 psi). The second produces 100 cubic meters per hour at a pressure of 0.124 MPa (440 gallons per minute with pressure 18 psi) with 5.9 kW input. The third supplies 115 cubic meters per hour at a pressure of 0.124 MPa (500 gallons per minute at 18 psi pressure) with input of 7.46 kW.

### 6.5.1.4 Component Variation Example: Threaded Fasteners

The threaded fastener is an example of parameterization of a component illustrating how the fundamental principles underlying component function can be scaled. The threaded fastener's function is to apply a force. They are commonly used to hold two or more things together. The basic function of connection by means of a locally axial force finds application across a wide range of technological systems.

Threaded fasteners are available in a vast range of variations ranging in size from almost invisibly small to surprisingly large. Figure 6.8 shows a threaded fastener used in a mechanical watch mechanism. The fastener is smaller than the letters on a coin. Figure 6.8 shows a structural bolt used in building steel frame assembly. Each bolt has

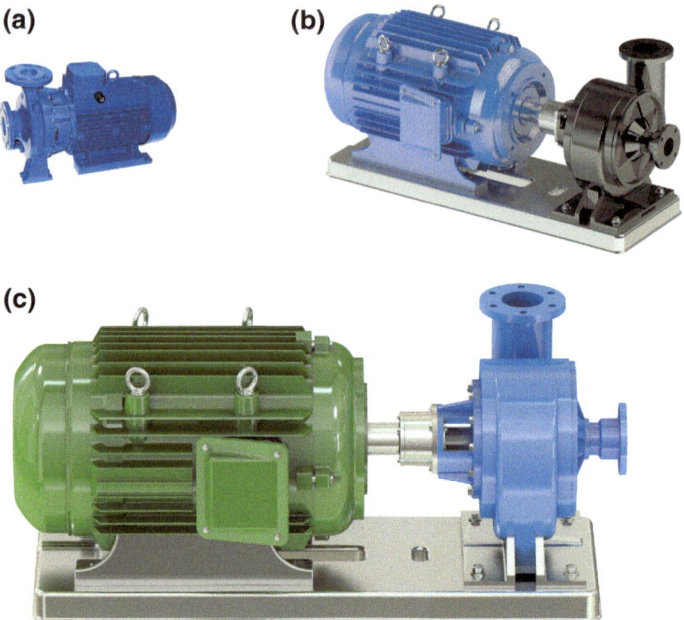

**Fig. 6.7** Centrifugal pumps: **a** (iStock.com/stefann11); **b** (iStock.com/AlexLMX); **c** (iStock.com/AlexLMX)

a mass of more than 10 kg (weight of more than 20 pounds). These examples show that the basic design of the component and the principle at work remain constant as the scale is changed.

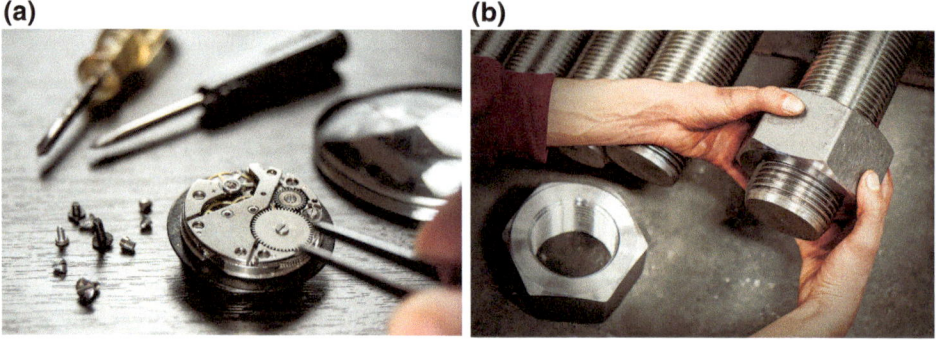

**Fig. 6.8** Large and small threaded fasteners: **a** watch with small fasteners (iStock.com/vefimov); **b** large industrial bolts (iStock.com/HAYKIRDI)

## 6.5.2   Parameterized Form Characteristics

Components are also parameterized to integrate effectively into the environment in which the component and system operate. Environment in this case refers to all aspects external to the component. Environment includes the physical conditions such as temperature and humidity. Environment includes interactions with the system user. Because interacting with humans is inherent in technology, components may also have to meet requirements due to the economic, social, and cultural environment into which the technological system is deployed. Systems might interact with other technological systems as well. All of these may create requirements on the system that influence the form or physical characteristics of a component. Groups of variations of a particular component are created to address different varieties of requirements.

Components are available in variations on the basic design. The function provided by a particular component may prove useful across a range of different technological systems. To help achieve the overall system function, features, or characteristics of the form of the component are made in different variations or otherwise parameterized. This parameterization enables the component to be utilized in a range of different requirements. The variations create components with features optimized for particular applications. The component features are varied to emphasize or provide additional functions or characteristics that are especially suited to a particular range of applications. Common characteristics leading to component variations are listed in Table 6.1.

**Table 6.1** Common characteristic feature attribute variations within component families

| Common component features |
| --- |
| Material properties |
| Size and dimensional characteristics |
| Weight |
| Expected lifetime |
| Appearance |
| Noise |
| Energy consumption |
| Maintenance requirements |
| Temperature range of operation |
| Resistance to corrosion or other chemicals |
| Color, texture, surface finish |
| Aesthetics and appearance |
| Toxicity or safety issues or features |
| Durability or stability over time |
| Mounting or attachment methods, other interface parameters |

The types of variations or parameterizations of a component occur across a spectrum from substantial to incidental. Component form may be varied to modify aspects such as the scale or magnitude of the fundamental transformations that occur in the component. Form variations may also exist in features such as color or styling. These form variations allow adapting components to systems which may vary in size, operating environment, application, or performance requirements. While some form features may be changed in component variations, the main physical principles at work in the component remain the same.

Component variations to better integrate with the needs of the system environment span a wide range of form characteristics. Table 6.1 lists some common features that may vary within families of the same component. Some common attributes that might vary include physical mounting and attachment configurations, temperature range of operation and resistance to water or weather conditions, color or aspects of aesthetics or appearance, expected durability and service life, safety features, and maintenance or service requirements, and cost.

### 6.5.2.1  Component Variation Example: Threaded Fasteners

The basic design of the threaded fastener is parameterized in characteristics such as diameter, length, number of threads, type of material, and head styles. These aspects of the design of the component are varied to match the varied requirements of different applications.

Figure 6.9 shows the same type and size threaded fastener made from different materials. The most common and lowest cost version of the fastener is made from plain steel. Steel provides high strength at low cost. A stainless version will not rust as does steel while providing comparable strength to steel but with higher cost. The aluminum version has reasonable strength with low weight and resistance to corrosion.

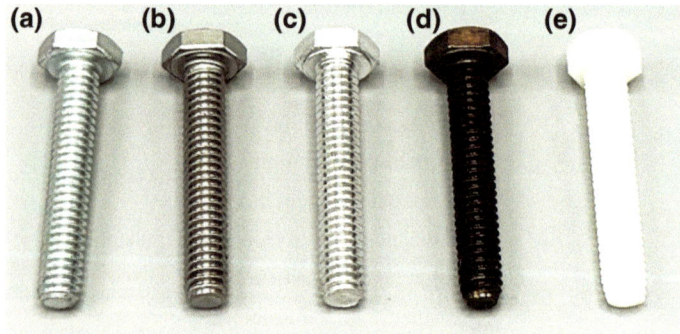

**Fig. 6.9**  Threaded fasteners of different materials: **a** steel; **b** stainless steel; **c** aluminum; **d** silicon bronze; **e** polymer nylon (photo by A. Krupczak)

For marine applications silicon bronze material resists corrosion from salt water. The fastener made of the polymer nylon might be used in applications requiring material that is not electrically conductive. Nylon is also more resistant to acids than metals.

All of these variations allow the fastener as a component to better meet the needs of the technological system in which it is used, however the basic underlying principle of operation is unchanged in all of these variations.

### 6.5.2.2  Component Variation Example: Integrated Circuit Packaging

The packaging of electronic integrated circuits is an example of how the form of a component can be modified to meet different requirements for size and interconnection to other system components.

An integrated circuit is a group of transistors and other electronic components that are combined into a single package. The integrated circuit carries out some well-defined function in converting electrical inputs into specific outputs. Examples include amplification, signal generation, and digital logic operations.

Figure 6.10 illustrates some of the different integrated packaging options for integrated circuits. For a given circuit all of the packaging options have the same inputs, outputs, and function. One option is Dual Inline Pin or DIP packaging. DIP packaging includes relatively large conductive pins on the package perimeter which typically connect to the rest of the system by fitting into holes in a circuit board. Another option is the Small Outline Integrated Circuit (SOIC) package. SOIC packaging is smaller than DIP packaging with interconnecting pins attaching to the rest of the system at connecting pads rather than holes. A third option is flat no-leads packaging such as Quad Flat No-leads packaging (QFN) shown in Fig. 6.10. No-leads packaging dispenses with metal leads for component interconnection and uses metal pads embedded in the perimeter of the packaging. No-leads packaging is the smallest size packaging for a given integrated circuit and allows a smaller overall device in space-critical systems such as mobile phones.

### 6.5.2.3  Component Variation Example: Electrical Switches

Electrical switches provide the function of actuating electric current in a technological system to control the flow of electrical energy. Several electrical switches are shown in Fig. 6.11. These switches all provide a basic ON/OFF function. The underlying principle in each switch is the same. Metallic conductors are put in contact to acuate current in the ON condition. In the OFF condition the conductors are separated.

Aspects of the form of each switch in Fig. 6.11 are specifically intended to help the switch to successfully integrate into a particular operational environment and fulfill specific system requirements. The foot switch is intended to be on a floor and actuated by a press from the user's foot. The housing is consequently of sturdy construction to withstand the relatively large force from the pressure from a foot.

The snap switch is frequently used as a sensor to detect the status of a door. The switch is often embedded in the door frames of appliances such as refrigerators and microwave

**Fig. 6.10** Examples of integrated circuit packaging: **a** coin for scale, **b** dual inline pin or DIP; **c** small outline integrated circuit or SOIC; **d** quad flat no-leads or QFN (photo by A. Krupczak)

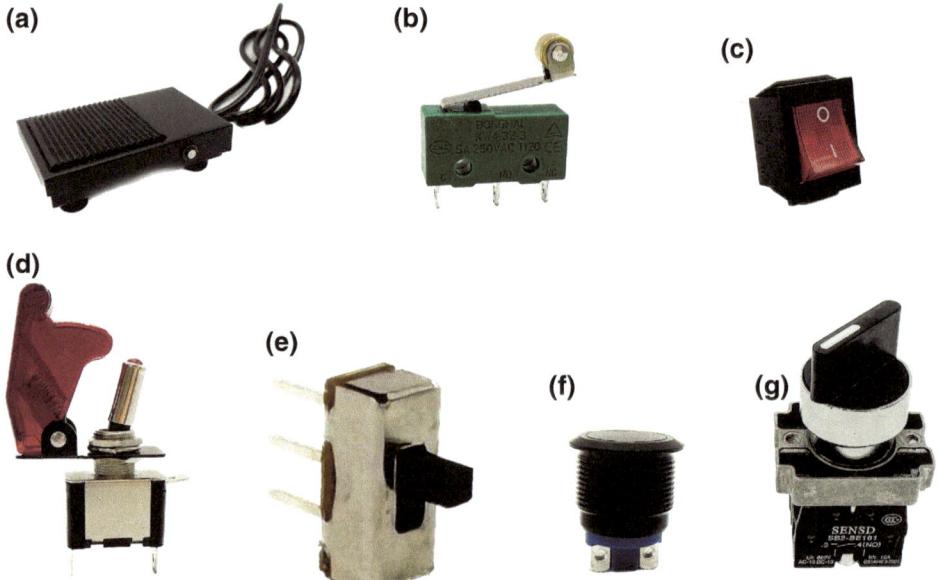

**Fig. 6.11** Different electrical switches: **a** foot switch; **b** snap switch; **c** rocker switch; **d** safety toggle switch; **e** slide switch; **f** push button switch; **g** rotational switch (photo by A. Krupczak)

ovens. The roller is pushed on by the door when it is closed. The ON/OFF status of the switch is used to control aspects of system operation such as turning on an interior light. The form features of this switch are optimized for small size, quiet operation, and ability to endure many thousands of ON/OFF cycles.

The rocker switch is optimized for use in small appliances such as coffee makers, space heaters, and air purifiers. The switch includes a light that is illuminated when the switch is in the ON position. The characteristics include relatively small size and resistance to moisture.

The safety switch is a toggle switch with a red hinged cover over the toggle. The cover helps to prevent accidental or inadvertent actuation of the switch. The user must lift the cover to enable access to the switch toggle. The additional step of lifting the cover helps to ensure more deliberate action by the operator.

The slide switch is a small, simple, and very inexpensive ON/OFF switch. It is intended to actuate relatively low currents. It finds application in very low-cost electronics and toys in which extremely low cost is a major requirement.

The pushbutton switch shown is used for ON/OFF function in large equipment in an industrial setting. The switch accommodates relatively large currents while being easy to activate by the user. The switch housing is heavy duty to endure rugged use.

Another industrial ON/OFF switch is shown in Fig. 6.11. The switch is rotated between ON and OFF positions. The rotational position of the lever helps the user to quickly identify the status of the switch from a distance. Rotational switches are also especially easy to assess status when viewing a bank of switches together.

### 6.5.2.4  Component Variation Example: Fluid Valves

Valves are used in fluid handling systems to control the flowrate of fluid at a particular point in the system. Three fluid handling valves are shown in Fig. 6.12. The valves all provide the same function. Features of the form of each valve are varied to align with the requirements that must be fulfilled to integrate successfully into different system environments.

The first valve is a general purpose ON/OFF valve. The valve is primarily made from brass and is suitable for common liquids such as water, oil, air, and inert gas. The valve

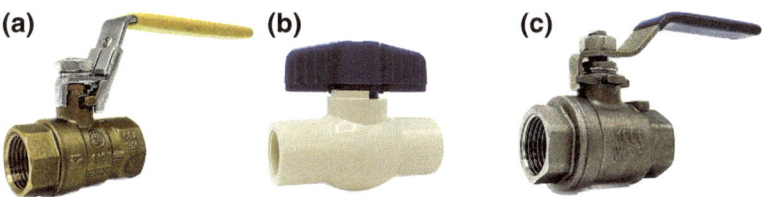

**Fig. 6.12** Different fluid valves: **a** brass general-purpose valve; **b** chlorinated polyvinyl chloride or CVPC valve; **c** stainless steel valve (photo by A. Krupczak)

can function at moderate temperatures below 90 °C (194 °F). The second valve is made from chlorinated polyvinyl chloride (CVPC). This material is well-suited as a shutoff valve for domestic hot and cold water. The third valve is constructed primarily from stainless steel with high corrosion resistance. This valve can be used in applications involving oil, acids, biodiesel, fuels, and alcohol.

### 6.5.2.5  Component Variation Example: Caster Wheels

It is not advisable to "reinvent the wheel" but the form characteristics of this component can be modified to better fulfill requirements placed on systems utilizing wheels. Figure 6.13 shows different types of caster wheels. For adoption in different systems with different needs and conditions different versions of the caster exist.

The first caster wheel utilizes soft rubber for the wheeled portion of the component. This soft rubber caster is useful for moving over carpet, ceramic, concrete, and wood floors. This component helps to meet requirements for low noise, durability, and smooth rolling.

The second caster has a plastic wheel. This implementation is low cost and light-duty for use on inexpensive household items and lightweight furniture. These wheels roll acceptably well in domestic spaces with carpet or wood flooring.

The caster with wide wheels is intended for office furniture such as office chairs. The plastic wheel is quiet and the extra-wide width distributes the load over a wider

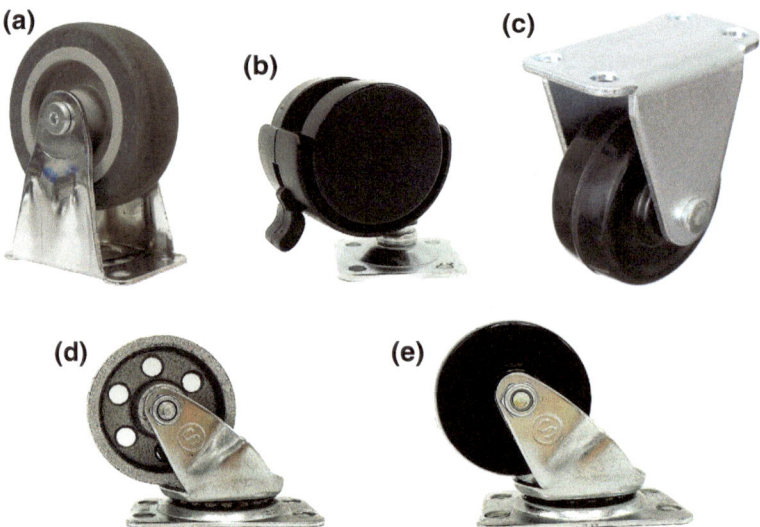

**Fig. 6.13**  Different types of caster wheels: **a** soft rubber caster wheel (iStock.com/Ekkaluck); **b** plastic caster wheel (photo by A. Krupczak); **c** wide caster wheel (iStock.com/Kevin Brine); **d** cast iron caster wheel (photo by A. Krupczak); **e** polyurethane caster wheel (photo by A. Krupczak)

contact area with the floor. This reduces the tendency of the wheel to sink into carpeting commonly encountered in offices.

The cast iron caster wheel is constructed entirely of metal. This caster wheel meets the requirements of infrequently moved objects on concrete and rough wood floors as might be found in industrial settings. The component maintains mobility in extreme heat and cold temperatures. The cast iron is resistant to oils and chemicals.

The industrial strength polyurethane caster is optimized for the requirements of smooth rolling and long service life. The construction enables easy and safe rolling of very heavy loads.

### 6.5.2.6 Component Variation Example: Electric Motor

Electric motors convert electrical energy into kinetic energy in the form of a rotating shaft. Almost anything that is electric and spins uses an electric motor. Electric motors have been developed that are optimized for a set of characteristics appropriate for particular applications.

Consider three examples of motors shown in Fig. 6.14. One motor is optimized for use in hairdryers. Besides the basic function of having an input of electrical energy and an output of a spinning shaft, this motor is optimized for low cost, light weight and low noise.

The second motor, used for an electric fan, is developed with characteristics that include a long life, high efficiency, ability to run for long periods without overheating, and low noise.

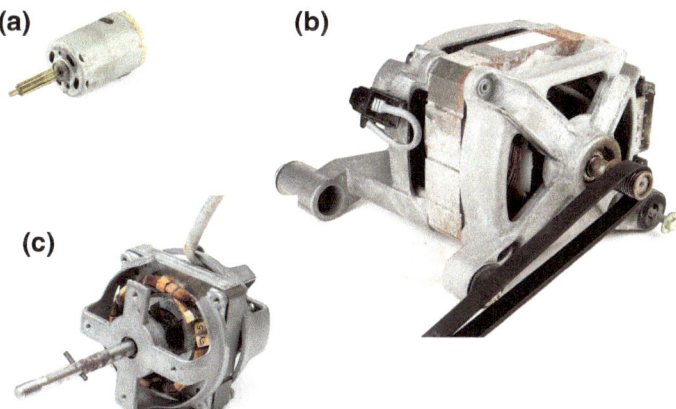

**Fig. 6.14** Examples of electric motors: **a** hair dryer motor (iStock.com/vitsirisukodom); **b** washing machine motor (iStock.com/Difydave); **c** fan motor (iStock.com/hideto111)

The electric motor used in larger home appliances like washing machines or clothes dryers is optimized to carry out the basic functions with added features of high torque (twisting force), low noise, and reliable operation with little maintenance.

## 6.6    Component Standardization

The development of parameterized components leads to the process of standardization by component manufacturers. Producers create components of predetermined standard configurations to address commonly encountered technological functions and system requirements. Established components become available with predetermined features and properties such as standard physical dimensions or levels of functional performance.

This standardization of technological system components is comparable to the situation for most mass-produced products. Consumer products provide a typical example. Everything from clothing to food is made available for purchase in predetermined sizes and quantities.

Standardized variations of the component facilitate adoption of components into new or different technological systems. Established components are available and able to be quickly utilized in new systems. Existing standard components help to reduce the cost and time needed to develop new technological systems. Standardization of components facilitates interchangeability as well as adoption of components into new or different technological systems.

One aspect of utilizing established standard components in a new system is finding a match between function and form requirements and the capabilities of an existing component. The performance level or magnitude of the component transformations and the form characteristics of an existing component many not be an exact match to the new system requirements. An evaluation must be made if an existing standard component has characteristics sufficiently close to ideal specifications to adequately meet all relevant requirements of the new system.

If the demands of a particular system warrant the expense, it is possible to develop or obtain components that are not standard but are customized to a particular application.

## 6.6.1    Manufacturer's Standards

Component manufacturers utilize standards. This allows interchangeability of components across different suppliers. This standardization means that in a technological system the design may be constrained to adopt a component that is available in a standard value of component parameters. System designers must be aware of relevant standards and how particular standard components may influence design choices.

While the existence of standards can limit the choices of the engineer, the use of standards generally increases the efficiency of the process of creating new systems. Established components, available in standard configurations, avoid the need to re-create an established component. Industry or government organizations establish standards and may certify compliance with standards by particular manufacturers.

The notations or descriptions of components often reflect standardization. The designation of the component includes some information relating to some of the parameters of the component. Standards can range from basic and minimal to multifaceted and detailed depending on the complexity of a particular component.

If the demands of a particular system warrant the expense it is possible to obtain components that are not standard but are customized to a particular application.

### 6.6.1.1  Example: Gears

Gears can function to transfer mechanical power from one rotating shaft to another. One gear exerts a force on another gear through the interlocking of gear teeth. Gears have long been used as a component in technological systems. The ancient Greeks used gears, Leonardo daVinci designed devices that used gears, and many systems appearing in the industrial revolution of the nineteenth century utilized gears.

Figure 6.15 shows a spur gear. Spur gears are circular and have the gear teeth arrayed around the circumference of the circle. Although the operating principle of gears is intuitive, the need for precise geometrical form characteristics is less widely appreciated. The gear teeth must be precisely shaped to mesh smoothly and securely with minimal friction. The existence of a supplier of pre-designed gears greatly reduces the effort needed to incorporate gears as a system component. A constraint for this convenience is utilizing a standard gear size and geometry already established by a manufacturer.

Table 6.2 provides an example of typical specifications for one supplier's standard small spur gears. Since gears physically connect with one another, data on dimensional parameters are particularly important in describing gear components.

**Fig. 6.15** Spur gear (iStock. com/lukelake)

**Table 6.2** Typical manufacturer gear specifications

| Gear pitch | Number of teeth | Gear pitch dia. | OD | Face Wd. | Overall Wd. | For shaft dia. | Material | Teeth heat treatment | Dia. | Wd. |
|---|---|---|---|---|---|---|---|---|---|---|
| 24 | 36 | 1 1/2″ | 1.58″ | 1/4″ | 0.625″ | 3/8″–5/8″ | 1144 Carbon steel | Not hardened | 1.125″ | 0.375″ |
| 24 | 48 | 2″ | 2.08″ | 1/4″ | 0.625″ | 3/8″–11/16″ | 1144 Carbon steel | Not hardened | 1.25″ | 0.375″ |
| 24 | 72 | 3″ | 3.08″ | 1/4″ | 0.75″ | 1/2″–13/16″ | 1144 Carbon steel | Not hardened | 1.375″ | 0.5″ |
| 20 | 24 | 1.2″ | 1.3″ | 3/8″ | 0.75″ | 3/8″–9/16″ | 1144 Carbon steel | Hardened | 1.063″ | 0.375″ |
| 20 | 24 | 1.2″ | 1.3″ | 3/8″ | 0.75″ | 3/8″–9/16″ | 1144 Carbon steel | Not hardened | 1.063″ | 0.375″ |
| 20 | 25 | 1 1/4″ | 1.35″ | 3/8″ | 0.75″ | 3/8″–5/8″ | 1144 Carbon steel | Not hardened | 1.109″ | 0.375″ |
| 20 | 32 | 1.6″ | 1.7″ | 3/8″ | 0.875″ | 3/8″–7/8″ | 1144 Carbon steel | Not Hardened | 1.438″ | 0.5″ |
| 20 | 36 | 1.8″ | 1.9″ | 3/8″ | 0.875″ | 3/8″–15/16″ | 1144 Carbon steel | Not hardened | 1.625″ | 0.5″ |
| 20 | 48 | 2.4″ | 2.5″ | 3/8″ | 0.875″ | 3/8″–1 1/4″ | 1144 Carbon steel | Not hardened | 2″ | 0.5″ |
| 20 | 60 | 3″ | 3.1″ | 3/8″ | 0.875″ | 3/8″–1 1/4″ | 1144 Carbon steel | Not hardened | 2″ | 0.5″ |
| 16 | 20 | 1 1/4″ | 1.38″ | 1/2″ | 0.938″ | 1/2″–9/16″ | 1144 Carbon steel | hardened | 1.063″ | 0.438″ |
| 16 | 20 | 1 1/4″ | 1.38″ | 1/2″ | 0.938″ | 1/2″–9/16″ | 1144 Carbon steel | Not hardened | 1.063″ | 0.438″ |

### 6.6.1.2 Example: Loudspeaker

The function of loudspeakers is to convert electrical energy into sound energy. The input is an electrical signal, and the output is sound. Speakers are a component found in some consumer electronics such as televisions, Bluetooth speakers, and soundbars. A selection

**Fig. 6.16** Audio loudspeakers (iStock.com/Aleksandr Kharitonov)

of standard speakers from one supplier are illustrated in Fig. 6.16. The supplier information about the standard available speakers provides specific data about the functional performance and form characteristics of the speakers.

Some of the information reported for these standard speakers addresses the characteristics of the speaker function converting an electrical signal input into a corresponding sound output. The input parameter provided is the maximum electrical power input measure in Watts (W). The output information supplied includes the output sound pressure level (maximum loudness) produced given in terms of decibels (dB). The other important output is the frequency or pitch of the output sound that the speaker is able to output. This is measured in Hertz (Hz).

The form parameters included in this instance center on characteristics relevant to connecting the speaker to other components. These are the physical dimensions and the electrical impedance of the speaker. Information about features common to all speakers in that group include cone type, magnet type, coating on metal surfaces to prevent rusting and type of interconnecting wires internal to the speaker (Table 6.3).

## 6.6.2 Industry-Wide Standardization

What are standards? The tendency of manufacturers to create technological components with pre-determined functional and form characteristics leads to the development of industry-wide standards for popular components. Industry standards are documents that describe important features of a product. Standards can specify aspects of the materials,

**Table 6.3** Audio loudspeaker available variations

| Dimensions (Dia. "xD") | Input (W) | | Impedance Ω | Output SPL (dB) | Frequency Range (Hz) | Magnet size (Dia. "xH") |
|---|---|---|---|---|---|---|
| | Nom. | Max. | | | | |
| $26 \times 0.9$ | 3.0 | 4.0 | 4 | 95 | 300–14,000 | $1.8 \times 0.5$ |
| $26 \times 0.10$ | 0.2 | 0.3 | 8 | 82 | 500–6000 | $1.3 \times 0.5$ |
| $26 \times 0.11$ | 0.2 | 0.3 | 8 | 97 | 500–6000 | $1.3 \times 0.4$ |
| $26 \times 0.12$ | 6.0 | 8.0 | 8 | 97 | 800–10,000 | $1.8 \times 0.5$ |
| $26 \times 0.13$ | 0.2 | 0.4 | 16 | 97 | 330–8000 | $1.3 \times 0.5$ |
| $26 \times 0.14$ | 2.0 | 3.0 | 8 | 85 | 300–10,000 | $1.8 \times 0.6$ |

physical dimensions, material properties, and functional performance characteristics of a component.

Standards make it possible for multiple manufacturers to produce components meeting the same set of requirements, increasing choice for consumers. Many thousands of standards exist worldwide. Standards address components ranging from basic electrical outlets to integrated circuits used by mobile phones. By offering components that meet relevant standards manufacturers ensure that their products are consistent, compatible, and safe.

### 6.6.2.1  Standards Creation

National and International organizations oversee the development of technological standards. Major international agencies creating standards include the International Organization for Standardization (ISO) and the International Electrotechnical Commission (IEC). In the United States engineering and technical professional organizations carry out standards development as part of the organization's activities. Examples include the American Society for Testing and Materials (ASTM), American Society of Mechanical Engineers (ASME), Institute of Electrical and Electronic Engineers (IEEE) and the Society of Automotive Engineers (SAE). In some instances, standards originally developed in one country become adopted internationally. Many standards originally established by The German Institute of Standardization (DIN) are widely used internationally. In some cases, more than one standard may exist for a component. This is especially the case for widely used components.

### 6.6.2.2  Standard Naming

Frequently, standardization of components is accompanied by standardized naming or identification systems. The standardized naming systems facilitate use of components. Manufacturers, suppliers, and end users can all use the standardized name to refer to the same component. This reduces problems that occur in trying to describe and specify particular features or aspects of components. Working directly with these components involves learning to use the associated standardized labels and designations.

Because of the large number and variety of standardized components and the needs of different applications it is impossible to generalize. However standard component naming systems usually follow a logical pattern. Resources exist to explain the standardized naming systems of particular component families and these are readily learned as needed. These become more familiar with use.

### 6.6.2.3   Standardization Example: I-beam

An I-beam is a component used to transfer loads in structures. The name "I-beam" refers to the characteristic cross section of the beam. I-beams provide structural rigidity with relatively low weight and material usage (Fig. 6.17).

I-beams are a common component used worldwide and multiple standards exist. These include the ASTM International—ASTM A6/A6M-21 Standard, the German DIN 1025 standard, and the AS/NZS 3679.1—Australia and New Zealand standard. These standards specify the dimensions, sectional properties, and masses of a group of I-beams.

Table 6.4 shows some of the I-beam sizes in American Standard Beams ASTM A6 standard. The standard designation of the I-beam size contains information about some parameters of the component. I-beams are an example of a component for which the code used to describe a particular standard size conveys some basic data about the component. In the ASTM A6 system the first number is the height of the beam in inches and the second number is the weight in pounds per foot of length. The S3 × 5.7 size then has a height of 3 inches and weighs 5.7 pounds per foot. The prefix S indicates that this is

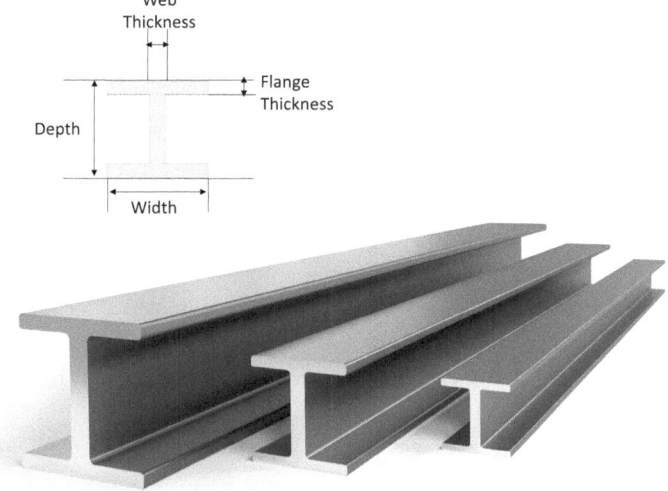

**Fig. 6.17**   Structural I-beams (iStock.com/bbsferrari)

**Table 6.4**  I-beam sizes in American standard beams ASTM A6 standard

| Designation imperial | Dimensions | | | |
|---|---|---|---|---|
| | Depth | Width | Web thickness | Weight |
| (in × lb/ft) | -h- | -w- | -s- | (lbf/ft) |
| S 24 × 121 | 24.5 | 8.05 | 0.8 | 121.00 |
| S 24 × 106 | 24.5 | 7.78 | 0.62 | 106.00 |
| S 24 × 100 | 24.0 | 7.425 | 0.745 | 100.00 |
| S 20 × 96 | 20.3 | 7.20 | 0.8 | 96.00 |
| S 20 × 86 | 20.3 | 7.06 | 0.66 | 86.00 |
| S 20 × 75 | 20.0 | 6.385 | 0.635 | 75.00 |
| S 15 × 50 | 15.0 | 5.64 | 0.55 | 50.00 |
| S 15 × 42.9 | 15.0 | 5.501 | 0.411 | 42.90 |
| S 8 × 23 | 8.0 | 4.171 | 0.441 | 23.00 |
| S 8 × 18.4 | 8.0 | 4.001 | 0.271 | 18.40 |
| S 7 × 20 | 7.0 | 3.86 | 0.45 | 20.00 |
| S 7 × 15.3 | 7.0 | 3.662 | 0.252 | 15.30 |
| S 5 × 14.75 | 5.0 | 3.284 | 0.494 | 14.75 |
| S 5 × 10 | 5.0 | 3.004 | 0.214 | 10.00 |
| S 4 × 9.5 | 4.0 | 2.796 | 0.326 | 9.50 |
| S 4 × 7.7 | 4.0 | 2.663 | 0.193 | 7.70 |
| S 3 × 7.5 | 3.0 | 2.509 | 0.349 | 7.50 |
| S 3 × 5.7 | 3.0 | 2.33 | 0.17 | 5.70 |

a "standard" I-Beam. There are also "wide flange" I-beams. These standard beams are available in heights ranging from 3 to 24 inches.

The German DIN 1025 standard beams are shown in Table 6.5. This system designates the beams with an "I" followed by a number. The number is the height of the beam in millimeters. The smallest beam in this system is "I 80" with a height of 80 mm. The largest "I 500" has a height of 500 mm or one-half of a meter.

### 6.6.2.4  Standardization Example: Electrical Resistors

Electrical resistors cause a decrease in voltage between their input and output ends. The amount of voltage change is proportional to the amount of current in the resistor and the amount of the resistance in the component. Resistors can serve the function of controlling the voltage or current at a particular point in an electrical system. The quantity of electrical resistance is measured in units called "ohms." Resistors are a commonly used component and a variety of types are available. One widely used type is a small cylinder with two wire leads. One wire is the input; the other wire is the output. A resistor is shown in Fig. 6.18.

**Table 6.5**   DIN I-beam sizes

| Designation Depth (mm) | Dimensions width (mm) | Static parameters | | |
|---|---|---|---|---|
| | | Web thickness (mm) | Sectional area (cm$^2$) | Weight (kg/m) |
| h | w | s | | |
| I 80 | 80 | 42 | 3.9 | 7.58 |
| I 100 | 100 | 50 | 4.5 | 10.6 |
| I 120 | 120 | 58 | 5.1 | 14.2 |
| I 140 | 140 | 66 | 5.7 | 18.3 |
| I 160 | 160 | 74 | 6.3 | 22.8 |
| I 180 | 180 | 82 | 6.9 | 27.9 |
| I 200 | 200 | 90 | 7.5 | 33.5 |
| I 220 | 220 | 98 | 8.1 | 39.6 |
| I 240 | 240 | 106 | 8.7 | 46.1 |
| I 260 | 260 | 113 | 9.4 | 53.4 |
| I 300 | 300 | 125 | 10.8 | 69.1 |
| I 340 | 340 | 137 | 12.2 | 86.8 |
| I 360 | 360 | 143 | 13 | 97.1 |
| I 400 | 400 | 155 | 14.4 | 118 |
| I 450 | 450 | 170 | 16.2 | 147 |
| I 500 | 500 | 185 | 18 | 180 |

Variations in the material composition determine the electrical resistance between the two wires.

A designer of a system seeking to control current or voltage using a resistor does not have unlimited choice for a value of resistance to use. A convenient approach suitable for

**Fig. 6.18**   Electrical resistor: **a** penny (iStock.com/Andrey_KZ); **b** resistor (iStock.com/Gueholl)

**(a)**

**(b)**

**Table 6.6** Standard resistor values

| Resistor | Resistance | | | | |
|---|---|---|---|---|---|
| | Ohms | | Kohms | | |
| | 10 | 100 | 1 | 10 | 100 |
| | 12 | 120 | 1.2 | 12 | 120 |
| | 15 | 150 | 1.5 | 15 | 150 |
| | 18 | 180 | 1.8 | 18 | 180 |
| | 22 | 220 | 2.2 | 22 | 220 |
| | 27 | 270 | 2.7 | 27 | 270 |
| | 33 | 330 | 3.3 | 33 | 330 |
| | 39 | 390 | 3.9 | 39 | 390 |
| | 47 | 470 | 4.7 | 47 | 470 |
| | 56 | 560 | 5.6 | 56 | 560 |
| | 68 | 680 | 6.8 | 68 | 680 |
| | 82 | 820 | 8.2 | 82 | 820 |

many applications is to use standard value resistors. Table 6.6 shows some of the standard value resistors available from a supplier. Only a portion of the available list is shown since there are hundreds of resistors available.

In creating a system utilizing this standard component, the designer would need to select one of the available values of resistance. However, since resistors can be combined, the option exists to use multiple standard values to achieve a specific value if desired. For example, there is no 20 $\Omega$ unit available but 20 could be achieved by combining two 10 $\Omega$ resistors.

Several international standards exist for this type of resistor. The standards specify both resistance and tolerance. The tolerance is the percentage variation allowed from the nominal value. For example, a 100 $\Omega$ resistor with a 5% tolerance is allowed to have a resistance between 95 and 105 ohms.

The standard *IEC 60,062:2016* was developed for indicating the resistance value using colored bands around the resistor. There is an international standard based on the color and position of the bands. The color-coding standard used depends on the resistor precision. A commonly encountered standard uses four bands. The first three colors are used to designate the resistance value and the fourth band indicates the tolerance. Figure 6.19 shows this color coding. The first two bands are the first two digits of the resistance. The third band is a power of 10 multiplier. The fourth band indicates the tolerance.

For example, brown-black-red-gold is $10 \times 100 = 1000$ with 5% tolerance.

Resistor Color Code Calculator

| Resisistor | Band 1 | Band 2 | Band 3 | Band 4 |
|---|---|---|---|---|
| Color | 1st | 2nd | Multiplier | Tolerance (+/-) |
| Black | | 0 | $\times 10^0$ | |
| Brown | 1 | 1 | $\times 10^1$ | 1% |
| Red | 2 | 2 | $\times 10^2$ | 2% |
| Orange | 3 | 3 | $\times 10^3$ | 0.05% |
| Yellow | 4 | 4 | $\times 10^4$ | 0.02% |
| Green | 5 | 5 | $\times 10^5$ | 0.50% |
| Blue | 6 | 6 | $\times 10^6$ | 0.25% |
| Purple | 7 | 7 | $\times 10^7$ | 0% |
| Gray | 8 | 8 | $\times 10^8$ | 0.01% |
| White | 9 | 9 | $\times 10^9$ | |
| Gold | | | $\times 10^{-1}$ | 5% |
| Silver | | | $\times 10^{-2}$ | 10% |
| Pink | | | $\times 10^{-3}$ | |

**Fig. 6.19**  Resistor color code (illustration by author)

## 6.7    Component Data Sheets

Component manufacturers enable component adoption by providing information about available components and their properties relevant to incorporation into systems. Component parameters of frequent interest include performance characteristics, operating ranges, interconnection parameters, and physical dimensions. Manufacturers' component information facilitates achieving function and form requirements for component incorporation into a system. Manufacturers' product information or "data sheets" are usually a few compact pages, however, they can be longer in some cases.

The intended audience for component information is engineers and system designers that are potential adopters of the component. These engineers and designers when creating a new, or modifying an existing technological system may have requirements that can be fulfilled by the component. Technical details and quantitative specifications of interest to a design engineer are presented. However, this type of product overview summary is generally not the same as an operating or user manual. Data sheets are typically more succinct than a user manual. Serving to provide information relevant to decision making by a specialized audience of engineers and technical people, product information is more detailed and technically oriented than marketing or advertising, but less comprehensive than an operating manual.

Component parameters of frequent interest include performance characteristics, operating ranges, interconnection parameters, and physical dimensions. Table 6.7 lists some types of component information that manufacturers include in data sheets. The information supplied addresses the important nature of components including outputs, required inputs, standards met or used in the component, significant features or variations, characteristics affecting interconnection or interface with other system components. Not all information is included in every data sheet. Usually, the component manufacturer provides information most relevant to those that may be selecting this component. The amount of information and extent of the data sheet varies with the complexity of the component and the amount of information that might be required to successfully integrate the component into a system.

There is no uniform format for product datasheets. The format for conveying component parameterization varies depending on type of component, industry, technical background of the target audience, and level of detail considered useful to the intended audience. Manufacturers and suppliers include the types of specifications and details most useful for engineers and system designers seeking a component to provide a function and fulfill requirements.

**Table 6.7** Types of information included in data sheets

| Component information |
| --- |
| Product description |
| Notable features |
| Examples of applications |
| Environmental limits of operation |
| Physical dimensions |
| Packaging options |
| Standards that apply to the component |
| Performance specifications |
| Limits on minimum and maximum operating parameters |
| Material specifications |
| Safety information |
| Recommended installation procedures |
| Numbering or labeling system used |
| Tolerances or ranges for critical parameters |
| Technical data |
| Specifications for constituent elements |
| Variations available |
| Relevant patents |

Paradoxically, often some types of generic or basic aspects of the component are not described because it is assumed that the potential user is already familiar with the component function. Component data sheets usually do not include much information about underlying physical principles pertaining to how a particular component works. Space restrictions in working within a one- or two-page limit, lead to an emphasis on information most relevant to selecting and using the component.

Generally, cost information is not included on the data sheet itself. This is because prices can vary with time, quantity purchased, or other incentives offered by the supplier. The data sheet can describe the technical details of a component that remain permanent as purchase cost can change. Pricing information is often supplied separately from technical specifications.

Data sheets usually include more information than any one user might need. Using data sheets is often a matter of selectively finding specific information needed for a particular application and ignoring specifications that are not relevant to the user's application. Finding information on data sheets can be challenging since the format of the information may be unfamiliar.

Photovoltaic modules are a central component in both utility-scale and domestic photovoltaic systems as shown in Fig. 6.20.

A manufacturer's datasheet for a solar photovoltaic module provides an example of a typical product datasheet. The datasheet is shown in Figs. 6.21 and 6.22. The manufacturer is Mission Solar of San Antonio, Texas USA.

The first page of the datasheet provides an image of the product and the highly important information about the rated electric power produced of 395 W. Also included is the expected operating life of 25 years and details of resistance to extreme weather. Details

**Fig. 6.20** Photovoltaic solar panels being installed on a roof (iStock.com/anatoliy_gleb)

**Fig. 6.21** Mission Solar photovoltaic module data front sheet (used with permission of Mission Solar)

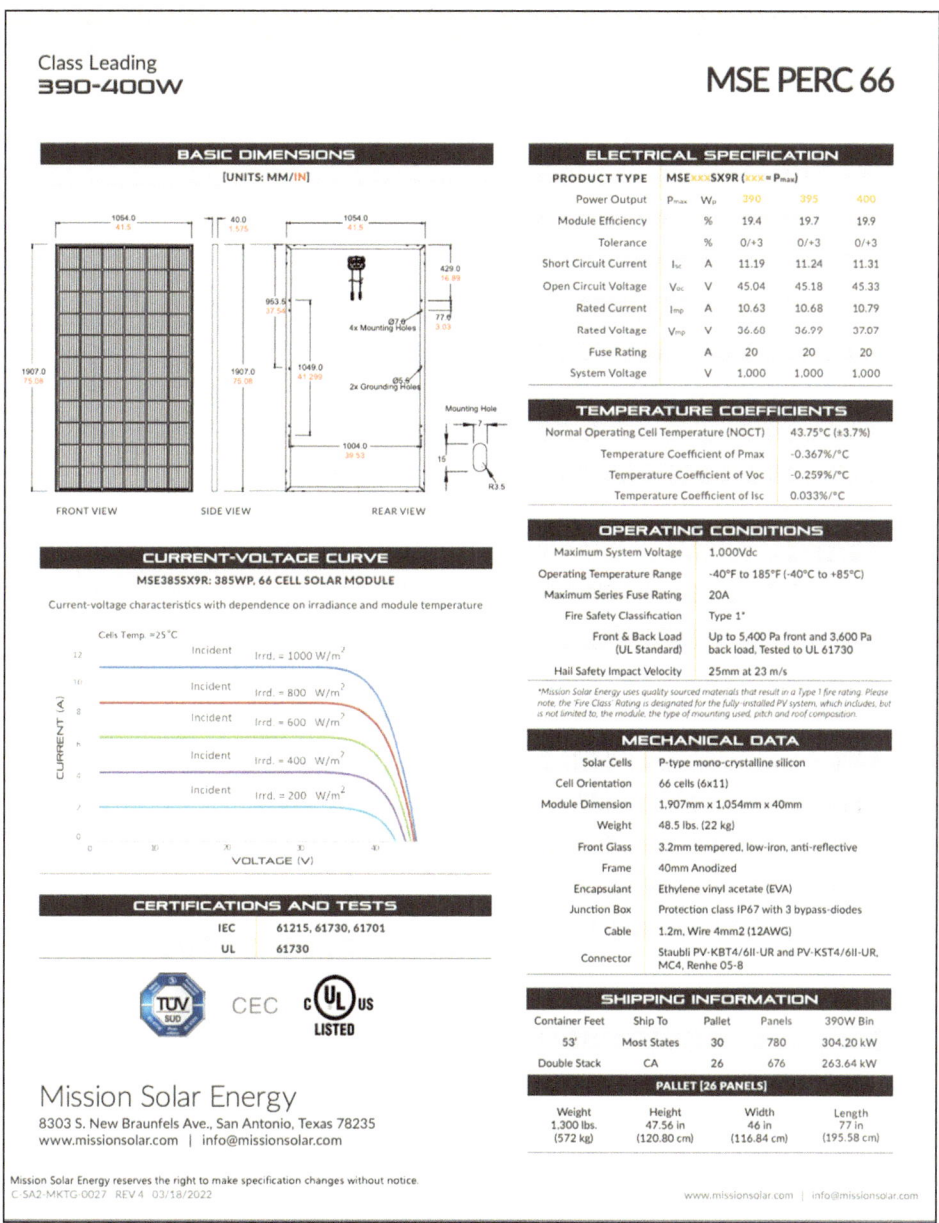

**Fig. 6.22** Mission Solar photovoltaic module data back sheet (used with permission of Mission Solar)

of compliance with specific relevant codes and certifications for this product are also included.

The second page of the datasheet provides specific physical dimensions of the module. Important electrical characteristics and mechanical data for the module are described. Current–voltage curves are shown at different levels of incident irradiance. This information characterizes the module output at different levels of solar intensity. Allowed ranges of operating conditions such as maximum operating temperature and maximum system voltage are specified. The datasheet also includes shipping information such as the number of panels per standard shipping pallet and container.

The datasheet does not include background information such as the underlying principles of how the panels convert solar energy into electricity. This is typical of datasheets; the emphasis is on the form and function details of a particular component. The datasheet aims to assist a potential user by providing information relevant to the integration of the product into the design of a system.

## 6.8    Component Vocabulary and Generic Component Descriptions

For describing, analyzing, and designing technological systems a component vocabulary is helpful. A vocabulary or working set of components is useful in interpreting technological systems. A challenge is a multitude of variations exist for most commercially available components. No single specific instance is a universal representation of a component.

It might be instructive to develop an analogy with biology and biological systems or ecosystems. Figure 6.23 shows four items. Which is the bird? In an ecosystem what do birds do? How do birds behave? In other words, if you watched a bird for a period of time, what types of action would you observe?

In responding to this question most people make use of their internal mental database associated with birds. This is a mental representation of characteristics and features associated with the idea of birds. If bird is in their vocabulary of animals, they have some ideas about birds and bird characteristics and behaviors.

The bird in the figure is not any specific bird but it is recognized by characteristics associated with birds. Is has a general shape common to many birds. It has features like a beak and eye common to birds. It has wings and feathers characteristic of birds.

Regarding behaviors in the environment in which it is found, a characteristic typical behavior of many birds is flying. Birds also make nests of found materials, eat bugs and seeds, and lay eggs. Generally, a vocabulary that includes birds would probably include these behaviors as associated with birds even though they are not specific to all birds.

A vocabulary of animals includes information about how to recognize them and expected behaviors in their ecosystem. A similar type of vocabulary of components is helpful in engaging technological systems. Figure 6.24 shows four technological components. Which of these is a centrifugal pump and what does it do?

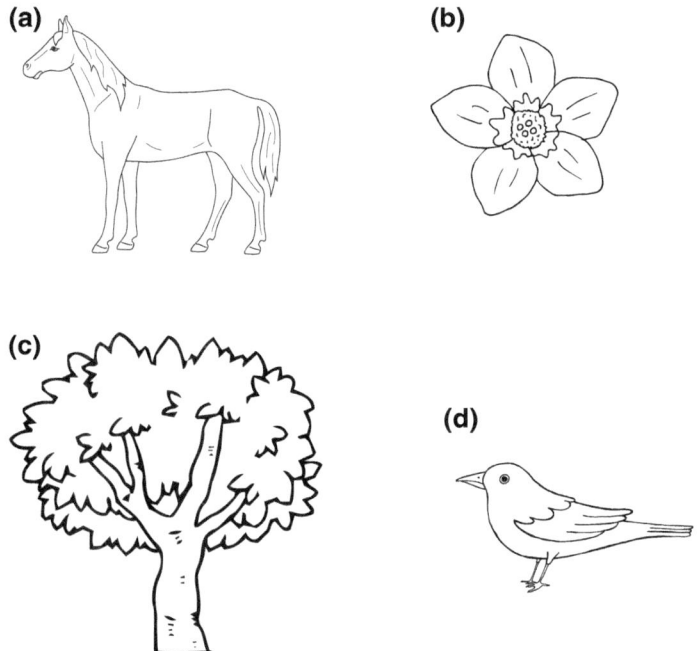

**Fig. 6.23** Four biological systems: **a** horse (iStock.com/Kositskaya Olga); **b** flower (iStock.com/Svetlana Soloveva); **c** tree (iStock.com/Murata Yuki); **d** bird (iStock.com/Irmun)

Identifying a particular technological component is similar to identifying a generic image of a bird. What are the characteristic features of the geometric form of the component? This process requires an internal mental database that includes a composite structure based on many similar but not identical variations of that component. Similarly, describing the behavior of the component, like the bird, requires a familiarity with the role of that component in technological systems.

An additional note, different people have a need for and develop different mental models of objects. Everyone's internal model or internal database of birds is not the same. Particularly, people that somehow need to or chose to interact with birds more frequently are likely to have a more detailed internal mental database of the characteristics and behaviors of birds, or certain birds. The same would apply to technological component vocabulary.

It can be helpful to suggest an approach for developing a brief overview or summary of essential information regarding a component. Following the analogy to biology, it is useful to have a generic internal mental representation or "component vocabulary" that makes it possible to identify or recognize a component in its environment or technological ecosystem and describe its behavior or function in this environment. The diverse information available from numerous sources pertinent to components suggests that a deliberate

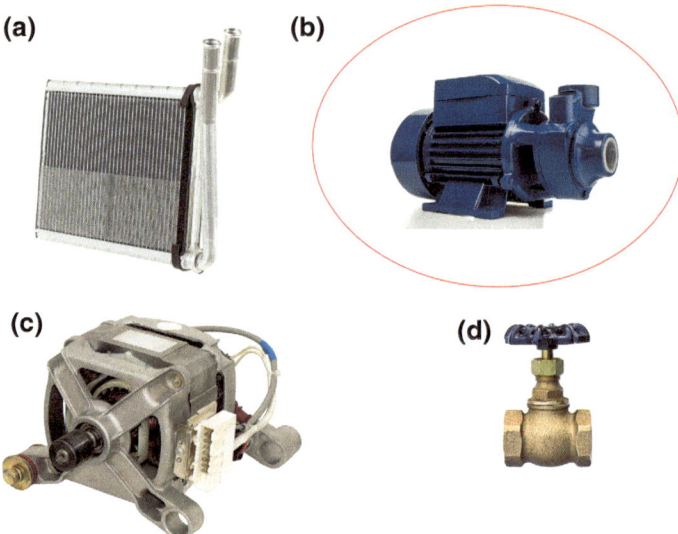

**Fig. 6.24** Four technological components: **a** car radiator (iStock.com/Stason4ic); **b** centrifugal pump (iStock.com/VladK213); **c** washing machine motor (iStock.com/ekipaj); **d** globe valve (iSt ock.com/cameracantabile)

approach to summarizing component characteristics might be an efficient organizer. The purpose of the summary is to collect characteristic information in a "generic datasheet" for that component.

A generic datasheet or component synopsis can provide a brief general overview of a component. Data sheets are prepared by manufacturers to describe details about their own specific products. A generic datasheet for the component might describe the component more generally.

Similarly to identifying animals, a generic datasheet component or synopsis can serve as a "field guide" of relevant or important information about the component. The synopsis is an individual reference guide about a particular component. The content and level of detail will reflect individual needs just as an individual's vocabulary of words reflects individual experience and needs.

What might be included in a generic datasheet for a component? The synopsis might contain some information also found on specific manufacturer's data sheets for that component but since data sheets are available it is not necessary to duplicate everything on a manufacturer's data sheet. The generic datasheet or component summary could be an overview of the component.

The component summary lists the intended function of the component. The emphasis is on intended because it is important not to lose sight of the fact that the form of the component could possibly be used for many different functions other than the original design intent.

How individuals construct and maintain mental representations of the physical world is beyond the scope of the present discussion, however, it might be efficient to identify aspects of a component summary that align with the description of technology as a system of interacting elements.

Table 6.8 is a list of some types of information relevant to a component overview summary or generic data sheet. Major categories of interest might include information about the function and form of the component along with a summary of the underlying principles or physical phenomena used in the component. Typical applications or systems that frequently utilize the component could be included. A summary could choose to highlight particular features or notable requirements relating to form and function provided by the component.

**Table 6.8**  Representative contents for a generic component datasheet

- Function
- – Component function
- – Inputs and outputs
- – Quantitative performance characteristics
- – Different modes of operation if relevant
- Common Applications
- Principles of Operation
- – Operating principles involved and major physical phenomena utilized
- – Relationship between form and function
- – Summary mathematical models
- Form
- – Distinguishing or recognizable features
- – Description of form characteristics
- – Typical dimensions and characteristics
- – Interconnection or fabrication information
- Other Information
- – Notably useful features
- – Limitations or potential weaknesses
- – Environmental or operational limits
- – Applicable standards
- – Variations available on the basic design

An example of a generic component datasheet for a centrifugal pump is included in the last section of this chapter.

## 6.9    Component Materials

Materials form the constituents from which components are fabricated. Materials are the various substances from which technological systems are constructed. The physical properties of these constituent materials determine the form characteristics of components. Component function is made possible by form properties of materials.

Materials are form without function or more precisely form that is not yet assigned a function. Materials represent form in its pure state. The properties of the form have not yet been modified, adapted, or assigned to a function. The material is not yet fashioned into a component to provide a subfunction in a technological system. Materials represent functional potential.

Examples of common materials are steel, concrete, silicon, polyethylene, copper, aluminum, and ABS plastic. Some common materials are shown in Fig. 6.25. Materials have form properties such as density, yield strength, electrical conductivity, color, elasticity, and melting point that facilitate components fabricated from these materials to fulfill or carry out a particular function. Similarly to commercially supplied components, materials are available in particular sizes and compositions. Standards exist for the composition and testing of materials.

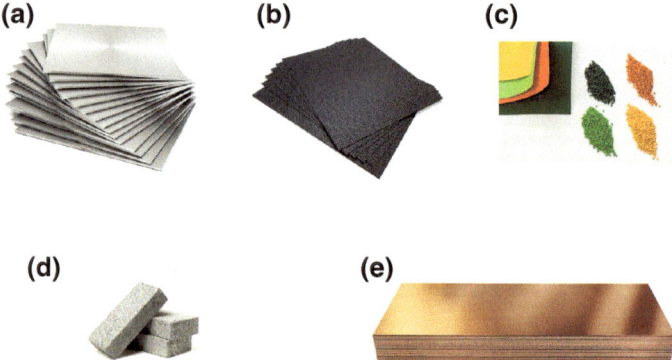

**Fig. 6.25** Examples of materials: **a** aluminum (iStock.com/adventtr); **b** carbon fiber (iStock.com/coddy); **c** ABS plastic (iStock.com/Irina Vodneva); **d** concrete (iStock.com/chengyuzheng); **e** copper (iStock.com/SimoneN)

### 6.9.1  Example Generic Datasheet for a Centrifugal Pump

The purpose of this section is to provide an example of a generic data sheet for a technological component. The generic datasheet summarizes the major aspects of component function and form in a general way. Manufactures' data sheets are necessarily specific to their particular versions of the component. This generic datasheet also provides background information about the underlying principles of component operation.

**Generic Data Sheet**

**Component Name: Centrifugal Pump**

**Function:**

*Component Function(s)*
Pump water or similar liquid.

Transfer mechanical energy into energy in form of hydraulic pressure of fluid.

Increase fluid pressure.

Move fluid.

*Inputs*
Flow of water or similar liquid at low pressure.

Mechanical power.

*Output*
Water or similar liquid with increased pressure.

Quantitative Performance Characteristics

Flowrate in volume per time (for example: gallons per minute or liters per hour).

Pressure increases between input output.

Input power used to accomplish transformation.

Different Modes of Operation (if Relevant)

The centrifugal pump has only one operating mode. It is either ON pumping fluid or OFF.

**Common Applications**
Centrifugal pumps are commonly used in systems for transfer of liquids. The pump increases the pressure of the liquid at the pump exit. This increased pressure is used

to push the liquid to higher elevations against the force of gravity or to overcome friction in pipe and hose.

Common applications include agricultural, building water and heating systems, food processing, chemical and pharmaceutical facilities, water treatment plants, power generation plants, oil and petroleum processing, and mining.

### Principles of Operation

*Operating Principles*

The main element of the centrifugal pump is the rotating impeller with curved vanes or blades as shown in Fig. 6.28. Figure 6.29 is a cut-away view of the pump showing the impeller located inside the pump casing.

Fluid enters the pump at the center eye of the rotating impeller. The entrance pipe at the center of the impeller can be seen in Figs. 6.28 and 6.29. The moving impeller vanes push on the liquid causing it to rotate along with the impeller. As the fluid rotates it experiences a centrifugal force pushing it away from the center and toward the perimeter of the impeller. As the fluid moves from the center of rotation its speed increases. High speed liquid moves into the volute which is the circular-shaped region that forms the outside of the pump casing around the impeller. When the fluid hits the wall, its speed decreases. The fluid pressure increases when the fast-moving fluid is slowed by hitting the wall. At this point the fluid's kinetic energy of motion is converted into energy in the form of fluid pressure. The high-pressure fluid exits at the discharge outlet located on the side of the casing.

The centrifugal pump requires an input of mechanical power. Some method is required to rotate the impeller. This is frequently accomplished using an electric motor. The output shaft of the motor is connected directly to the impeller shaft. Other drive systems are sometimes used including a belt drive or power from an internal combustion engine.

*Relationship between Form and Function*

The geometric shape of the impeller is a critical factor in the centrifugal pump. The vanes must form a ridge that can push on the water and cause it to rotate along with the impeller. The volume flowrate of liquid that can be pumped depends in part on the diameter of the impeller.

The donut-shaped volute volume around the perimeter of the pump tends to increase in size in the direction of the outlet. This helps to accommodate the increased flow of liquid without increasing velocity. The larger cross section of the volute donut accommodates a larger flow of liquid without causing the liquid to increase in speed. Faster moving fluid in this section would result in an undesired outcome of lower pressure.

*Mathematical Models*

The centrifugal force $F$ pushing the liquid toward the edge of the rotating impeller is given by:

$$F = m\omega^2 r$$

where: $m$ is the mass of an element of fluid, $\omega$ is the angular rotational velocity of the impeller, and $r$ is the distance of the fluid element from the impeller center. This shows that the force pushing the liquid out increases with distance from the center and rotation speed of the pump.

The conservation of energy is an underlying principle utilized in achieving function of the pump. For liquid such as a steady flow of water without significant friction or an elevation change, the conservation of energy can be expressed using the Bernoulli formulation.

$$\frac{p}{\rho} + \frac{V^2}{2} = constant$$

where $p$ is the liquid pressure, $r$ is the liquid density and $V$ is velocity. This expression of energy conservation applies to an element of fluid as it travels along its flow path. The rotating impeller and centrifugal force add energy to the fluid increasing the velocity $V$. When the liquid flows into the volute outer ring, $V$ decreases. However, energy is conserved so the Bernoulli expression shows that if $V$ decreases, the pressure $p$ must increase to keep the combination expressed in the left side of the equation equal to a constant. Energy in the form of velocity becomes energy in the form of increased pressure.

A mathematical model expressing a relationship between the input and output of the centrifugal pump is given by:

$$\dot{W} = Q\Delta p \frac{1}{\eta_p}$$

where $\dot{W}$ is the mechanical power input used to rotate the impeller. $Dp$ is the increase in liquid pressure at outlet compared to inlet. $Q$ is the flowrate of liquid through the pump measured as volume per time such as gallons per minute or liters per hour. The term $h_p$ is the pump efficiency. The efficiency will be a number less than 1, due to losses, all of the input power will not be converted to equivalent output power. This expression relates component outputs of liquid flow rate and pressure to the input of mechanical power used to rotate the impeller.

**Form**

*Distinguishing or recognizable features*

The housing of the pump is usually circular, and donut shaped with the inlet pump attached at the center of the circle and the outlet attached at the side along the circumference. These are distinctive form features through which the pump can be recognized. Centrifugal pumps in Figs. 6.26 and 6.27 display these distinctive characteristics.

**Fig. 6.26**   Centrifugal pump used for irrigation (iStock.com/Peerayot)

**Fig. 6.27**   Centrifugal pumps in an industrial application (iStock.com/LiuNian)

**Fig. 6.28**  Centrifugal pump impeller (iStock.com/kirill4mula)

**Fig. 6.29**  Cut-away view of pump showing impeller inside casing (iStock.com/Itsanan Sampun-tarat)

In many applications an electric motor to turn the impeller is directly attached at the pump housing. This is recognizable as a cylinder comparable in diameter to the size of the pump. Other drive configurations are possible in all cases attached to the impeller shaft.

*Significant form characteristics*
In addition to the characteristic geometry of the impeller and the volute necessary for the intended operation, the materials of the impeller and pump housing must be able to

endure continuous contact with the pumped liquid without experiencing detrimental rust or corrosion.

The shaft driving the impeller goes through the pump housing. This must be carefully sealed to prevent liquid from leaking from the pump.

Pumps usually rotate at high speed so attention must be given to reducing friction in the moving parts.

*Typical dimensions and characteristics*

A commonly used pump of the type shown in Fig. 6.30, can produce a flow rate of 33 gallons per minute (7500 L per hour) at pressure of 28 psi (198 kPa). It has physical dimensions of 12 inches length (300 mm) with a height of 8 inches (200 mm) and a weight of about 30 pounds (13.6 kg). An electric motor is used with a power input of 1/3 hp (250 W).

*Interconnection or fabrication information*

The pump connects to the system through piping at the inlet and outlet. Often flanges or pipe fittings are used to facilitate interconnection.

**Other Information**

*Notable useful features*

Centrifugal pumps are relatively uncomplicated to install and use. Many variations in size, fluids pumped, and operating environment are available. Some pumps can accommodate liquid that contains small amounts of solid material such as sand or silt.

**Fig. 6.30** Centrifugal pump used for irrigation (petchyai/Shutterstock.com)

Detailed system design including pumps is informed by the manufacturer's performance curves of pressure versus flowrate for a specific model pump.

*Limitations or potential weaknesses*

Most centrifugal pumps require that liquid be present in the inlet section for the pump to start. The pump requires a "prime" of liquid in the pump to start working. Centrifugal pumps are not "self-priming."

Low pressure in the liquid caused by the action of the impeller can result in formation of vapor bubbles that can damage the impeller. This is called "cavitation." Increasing inlet pressure reduces the possibility of cavitation.

Some centrifugal pumps cannot tolerate running without any liquid present due to friction and heat generated in "dry" operation.

*Environmental or operational limits*

Pumps often have a maximum environmental temperature at which they can operate. Pumps are also limited to specific types of fluids with which they can be used primarily due to potential corrosion or chemical reaction between the liquid and pump materials. Centrifugal pumps do not work well with liquid that contains significantly large entrained solid material such as rocks or gravel.

*Variations available on the basic design*

Some variations bend the inlet pipe so it is parallel to the outlet pipe (while still entering at the center eye of the impeller. This facilitates interconnection in some types of piping systems.

Different types of impellers are used to accommodate different amounts of solid material carried along with the fluid.

Applicable standards

ISO (International Organization for Standardization).

ISO 5199:2002 Technical specifications for centrifugal pumps.

This international standard is one of a set dealing with technical specifications of centrifugal pumps. Criteria for the selection of a pump for a certain application may include reliability, required operating life, operating conditions, environmental conditions, and local ambient conditions.

ASME (American Society of Mechanical Engineers).

ASME standard B73.1—2020 Specification for Horizontal End Suction Centrifugal Pumps for Chemical Process.

This standard is a design and specification standard that covers centrifugal pumps of horizontal, centerline discharge design. The standard includes dimensional interchangeability requirements and design features to facilitate installation and maintenance and to enhance reliability and safety pumps. The standard addresses interchangeability of pumps with respect to features such as mounting dimensions, size and location of suction and discharge, input shafts, and foundation bolt holes.

## Bibliography

American Society of Mechanical Engineers "ASME B73.1 - Spec. for Horizontal End Suction Centrifugal Pump - ASME." Accessed August 1, 2023. https://www.asme.org/codes-standards/find-codes-standards/b73-1-specification-horizontal-end-suction-centrifugal-pumps-chemical-process/2020/drm-enabled-pdf.

ASTM International, "Standard Specification for General Requirements for Rolled Structural Steel Bars, Plates, Shapes, and Sheet Piling." Accessed August 1, 2023. https://www.astm.org/a0006_a0006m-21.html.

Fox, Robert W., Alan T. McDonald, and John W. Mitchell. *Introduction to Fluid Mechanics*. John Wiley & Sons, 2020.

International Electrotechnical Commission (IEC), Marking codes for resistors and capacitors "IEC 60062:2016+AMD1:2019 CSV " Accessed August 1, 2023. https://webstore.iec.ch/publication/65655.

International Electrotechnical Commission (IEC), Preferred Number Series for Resistors and Capacitors, "IEC 60063:2015" Accessed August 1, 2023. https://webstore.iec.ch/publication/22011.

International Organization for Standardization "ISO 5199:2002(En), Technical Specifications for Centrifugal Pumps — Class II." Accessed August 1, 2023. https://www.iso.org/obp/ui/#iso:std:iso:5199:ed-2:v1:en.

Standards, European. "DIN 1025–5." https://www.en-standard.eu. Accessed August 1, 2023. https://www.en-standard.eu/din-1025-5-hot-rolled-i-and-h-sections-ipe-series-dimensions-mass-and-static-parameters/.

# System Interdependence

**7**

## 7.1     Chapter Overview

- Technological systems are interdependent on other systems including technological, socioeconomic, and natural systems.
- Technological systems generally depend on other technological systems for some necessary inputs. The outputs produced by a technological system can become inputs to another system.
- Technological systems must be constructed or manufactured using other systems that depend on resource development and transportation systems.
- The manufacturing facilities in which components are made from basic materials and where systems are assembled from components are themselves technological systems frequently incorporating special purpose automated machinery.
- All technological systems interact with the natural environment.
- Technological systems interact with the other systems that form the infrastructure of human societies.
- Technological systems are physical objects that become part of the circumstances in which humans live. New technological systems can cause changes in patterns of human behavior.
- Technological systems encompass social and cultural interactions. People and societies influence the development and operation of technological systems through the allocation of human capital and other resources needed for technological development, control of access to natural resources utilized by technological system operation, and establishment of the requirements these systems must fulfill to be allowed to operate within the socioeconomic system and accepted by consumers.

J. Krupczak, Jr., *Understanding Technological Systems*, Synthesis Lectures on Engineering, Science, and Technology, https://doi.org/10.1007/978-3-031-45441-7_7

## 7.2    Systems Feed Systems

Technological systems depend on other technological systems. Technological systems transform inputs into the desired outputs. The inputs and outputs can be classified as materials, energy, or information. Clearly, the inputs must come from some source. For most technological systems one or more critical inputs are provided by another technological system or systems during some modes of operation. Some inputs may be provided by the natural environment or human users, but most modern technological systems will not operate without the output of other technological systems in some phase of their operation.

Consider an automobile as an example. Let's assume it is an electric vehicle (EV). Figure 7.1 shows a block diagram representation of the overall system function of the EV. The modes of operation shown are the charging and use mode. The EV receives an input of electrical energy and produces an output of energy of motion. In an overall view the function of the EV is to convert electrical energy into energy of motion.

The electrical energy to charge the EV battery must be supplied by some source. If the system boundary is expanded the source of the electrical energy to supply the EV can be identified. This EV is being charged in a home garage, so the electrical energy is supplied by the house distribution system. Figure 7.2 provides a diagram of the major elements. The house is supplied from the utility grid which distributes the generated electricity to the users.

Various generation facilities provide electrical energy input to the grid. The electric grid is actually a complex series of interconnections, but for this analysis, consider it as a system in its main function to transport and distribute electrical energy from producers to consumers. The grid receives input from generation facilities and provides electrical energy output to consumers.

Currently in the United States the major types of generation are natural gas, coal, nuclear, hydro, wind, and solar. In an overall view, the input to the grid on a national basis is 38% supplied by natural gas, 22% coal, 19% nuclear, 9% wind, 6% hydro, and 3% from solar generation facilities.

# Electric Vehicle

Electrical        Energy of
energy            motion

**Fig. 7.1**   Electric vehicle overall function block diagram (iStock.com/3alexd)

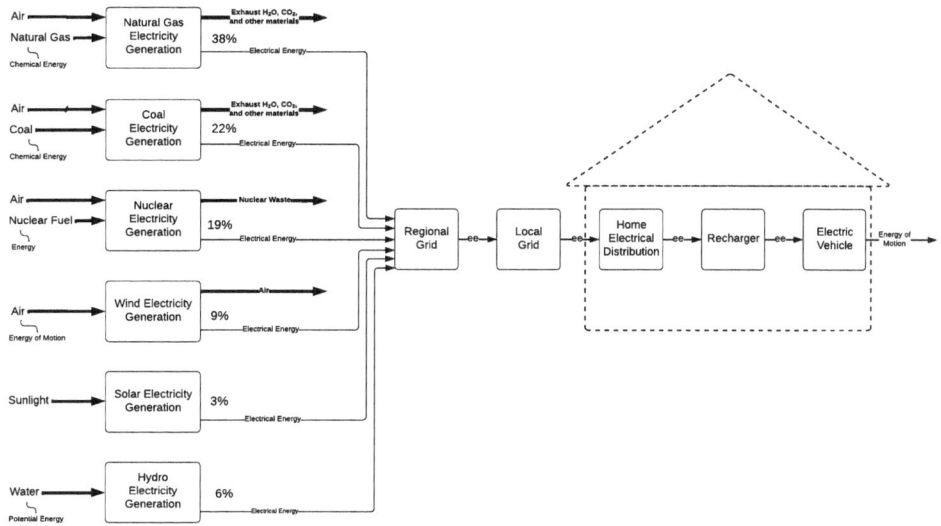

**Fig. 7.2**  Systems providing electricity to charge an electric vehicle (diagram by author)

Natural gas and coal generation convert chemical energy into electrical energy. Nuclear converts the nuclear energy into electrical energy. Hydro converts the energy of motion of the water into electrical energy. Solar converts solar energy into electrical energy. Finally, wind power converts the energy of motion of the wind into electrical energy. The electrical energy input to the EV is provided by these various energy conversion systems that supply the electrical energy distribution grid.

The EV system then is dependent upon the output of these functioning systems which in turn require inputs of natural gas, coal, nuclear fuel, wind, water, and solar energy.

## 7.3    Technological System Manufacturing

Technological systems must be manufactured by other systems that depend on resource development and transportation systems. Technological systems do not reproduce themselves. Modern technology is manufactured. The manufacturing process itself is a complex technological system. Manufacturing of end-product technological systems used by consumers relies on systems that produce constituent components. Component manufacturing in turn relies on input from systems producing materials. Basic materials are created by systems that extract and process material resources, transforming them into useable materials. Materials, components, and intermediate assemblies are transferred by transportation systems. All of the systems, from basic resource extraction to final manufacturing, require energy inputs for the transformations occurring at each processing stage.

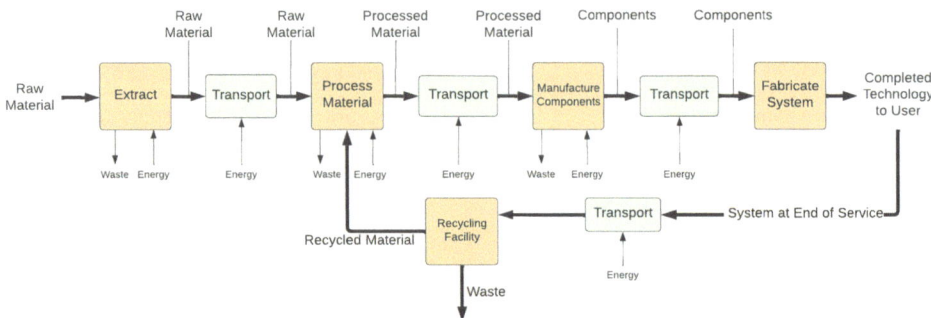

**Fig. 7.3**  Manufacturing systems (diagram by author)

Figure 7.3 is a diagram illustrating the essential features of the linked systems involved in manufacturing modern technological systems. The figure conveys the broad outlines of the process from extracted ore and unprocessed natural products. Of course, manufacturing and supply chains are more detailed and nuanced. In the figure, details have been omitted for clarity.

If the end of useful life is considered and recycling of some portion of the item occurs, then interaction with systems similar to that shown in Fig. 7.3 occurs. The technology at the end of its service life is conveyed by a transportation system to a recycling facility. Some portion of the system material is recycled. Recycling is usually at the level of base materials rather than functional components. For example, copper wire is usually recycled as copper metal rather than reused as wire. The recycling process requires energy input. The recycled material is shown symbolically reentering the manufacturing process at the material stage. Non-recycled materials leave the system as waste.

*Example: EV manufacturing and supply chain*
An overview of the manufacturing and supply chain for a typical electric vehicle demonstrates the extent to which the production of modern technological systems depend on complex supply and transportation systems. Figure 7.4 is a simplified representation of the systems that support a facility manufacturing EVs.

As illustrated in the system diagram of EV manufacturing shown in Fig. 7.4, the EV can be represented as having five major subsystems: electric motor, battery, body and suspension, wheels, and electronics.

The primary components of the electric motor are made from the metals copper, neodymium, and steel. These materials are assembled into the electric motor which is then transported to the vehicle final assembly site. The metals enter the system as ore minerals mined from the earth. The extracted ore is transported to sites where the metals are produced, refined and rendered into forms suitable for further fabrication steps into

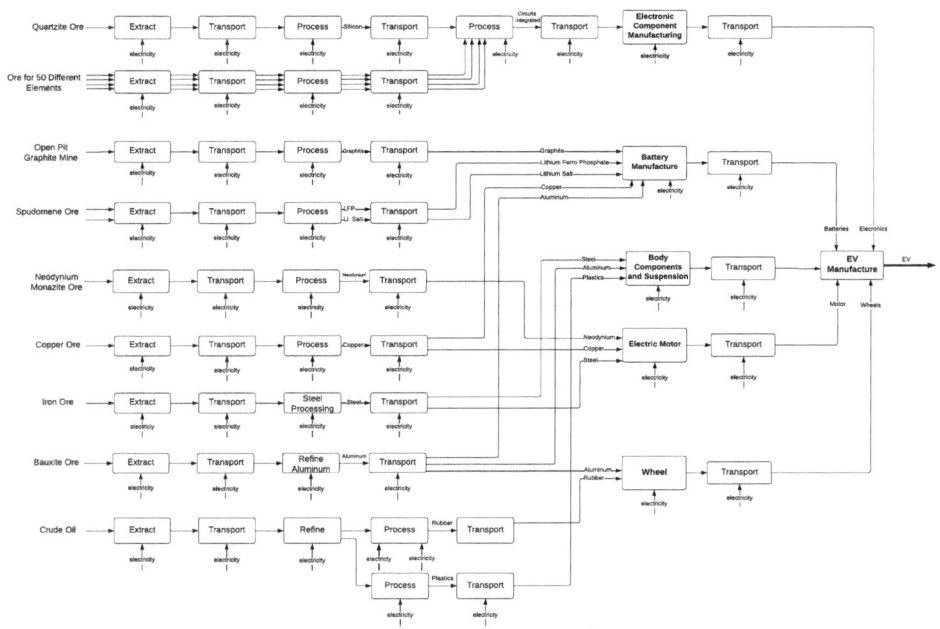

**Fig. 7.4**  Electric vehicle manufacturing (diagram by author)

components of the electric motors. Significant energy inputs are needed during all steps along with transportation of the materials between different locations.

Batteries are a major component of an electric vehicle. Batteries account for approximately 25% of the weight of EVs. Several battery chemistries exist but lithium is a main constituent in most EV batteries. Lithium ferro phosphate (LFP) is one common EV battery. Some of the LFP battery's primary components are made from lithium ferro phosphate, lithium salts, graphite, copper, and aluminum. These materials are at the end of supply chains that begin with spodumene lithium ore, lithium salar brine, bauxite, and open pit graphite mines (Fig. 7.5).

The vehicle body and suspension consist of components made of mostly aluminum and steel. Aluminum is derived from bauxite ore. Steel is primarily iron. The body and interior also include plastics and synthetics derived from crude oil.

Wheels are made from aluminum rims and rubber tires. Most rubber used in tires is synthetic. Synthetic rubber is made from petrochemicals derived from crude oil.

All modern vehicles utilize a significant amount of electronics. In the EV, electronics are used for the purpose of battery discharge and charge control, motor speed control, driver displays, navigation, braking control, lighting and signaling, and passenger climate system control. The electronics are at the end of a long and complex series of systems responsible for electronic components, integrated circuits, and intermediate assemblies.

**Fig. 7.5**  Lithium mining: **a** Greenbushes Lithium Mine, Western Australia (iStock.com/Zambez iShark); **b** Silver Peak Lithium Mine, Nevada, USA (iStock.com/simonkr)

Electronic systems utilize 50 of the 92 naturally occurring elements including silicon, germanium, copper, tin, bismuth, tantalum, and gold. Sources for diverse elements are mines dispersed throughout the world. Extraction, processing, assembly, and transportation occur globally with significant inputs of energy and materials throughout the process.

The example of the systems that support the manufacture of an EV demonstrates that the manufacture of modern technological systems relies on a diverse network of systems that create the materials and components from which end-product systems are constructed.

## 7.4    Systems Make Systems

Manufacturing and supply chains depend on systems that support the flow of materials from naturally occurring substances to basic materials to components and intermediate subsystems and final assembly. Manufacturing facilities, in which components are made from basic materials and systems are assembled from components, are themselves technological systems of special purpose automated machinery. Human operation and input do occur, but modern manufacturing primarily relies on automation to carry out most of the fabrication and assembly operations. Technological systems are used to make other technological systems.

Figure 7.6 is an image of a modern automotive assembly facility. Technological systems carry out the majority of the operations transforming the input materials into completed products.

**Fig. 7.6**   Vehicle assembly facility (iStock.com/Traimak_Ivan)

## 7.5   Systems Govern Systems

A system is a group of interacting elements, forming a network, to achieve a common purpose.

A system does not have to be a technological object. Major systems characterizing modern societies include systems of government, commerce systems, judicial systems, health care systems, education systems, and social systems. Technological systems become part of the physical objects that support human existence. In addition to requiring inputs and interactions with other technological systems, interactions occur with the other systems that form the infrastructure of modern societies.

Technological systems as elements in larger human organizations depend on the activities of these overarching socioeconomic systems. Availability and access to inputs of material, energy, and information required for technological systems manufacturing, and operation are determined and controlled by the functioning of these socioeconomic systems.

The example of an EV demonstrates the interconnection between technological systems and human socioeconomic systems. Table 7.1 lists some of the systems upon which the design, creation, and use of the EV and most automobiles depend.

The EV could not come into existence and would not function for long without the operation of these coexisting socioeconomic and other technical systems. Drivers could not employ current use patterns without these systems. Figure 7.7 showing an EV on an isolated island conveys the reality that the vehicle itself would not operate for very long in the absence of other systems that support and influence this technology.

**Table 7.1**  Socioeconomic and other technical systems supporting an EV automobile

| | |
|---|---|
| Banking and car loan system | Repair and maintenance establishments |
| Vehicle registration and licensing | Repair parts supply and distribution |
| Automobile insurance companies | Parking facilities |
| Laws regulating motor vehicles and driving | Driver training programs |
| Police and court system to enforce driving and traffic laws | International trade agreements |
| Road and highway system infrastructure | International transportation network to transport automobile components |
| Road construction and repair programs | Manufacturers and supplier system |
| Tax system to support roadway infrastructure | Industry network for recycling end-of-service vehicles |
| Vehicle and highway safety organizations | Software development system for vehicle-related applications |
| Charging stations | Mobile phone system |
| Electric utility system | Internet system |
| Consumer advertising | Broadcast radio system |
| Dealership and distribution system | Manufacturer user support |

**Fig. 7.7**  Isolated EV without necessary supporting systems (This image was created with the assistance of DALL-E 2 by author)

## 7.6 Technological Systems and the Natural Environment

All technological systems interact with the natural environment. This often takes the form of exchanges of material or energy during different stages of the product cycle of a particular technological system.

Aspects of the product cycle have been mentioned earlier in relation to the manufacturing phase of a technological product. The product cycle describes a simplified version of the life cycle of most human-made products. The first phase is extraction of resources from the earth from which the product will be made. The second phase is manufacturing. Next the product is packaged and distributed to the end user. The product then goes through a use phase. After the use phase, the product is returned to the natural environment as waste. Figure 7.8 shows the product cycle.

Due to the conservation of mass, the product cycle illustrates that whatever mass of material is initially extracted from the natural environment, eventually returns to the environment although in an altered state.

In some cases, material may be taken from the environment at phases other than the initial extraction phase. For example, cooling water added to an engine during operation. Waste products from intermediate stages of manufacturing represents material returned to the environment.

Each phase of the cycle typically requires energy input that must be derived from the natural environment in some form. Energy is conserved so any energy extracted from the environment is returned. However, the returned energy is often in the form of heat. Recycling of materials may occur with some products. Recycling of material requires inputs of energy.

**Fig. 7.8** Product cycle (iSt ock.com/petovarga)

## 7.7    Socio-Technological Impacts

New technological systems cause changes in patterns of human behavior. New technological systems are tangible objects that change physical reality as experienced by humans. When new technological systems are introduced, the physical world is changed in some way. Humans and human societies are adept at adapting to whatever physical environment in which they find themselves. Throughout history humans have proven to be creative and flexible in utilizing the resources of their physical environment to solve the problems of everyday life. It should therefore be expected that when new technological systems appear, changing the nature of physical reality in some way, that humans will adapt and adjust in some way to these new circumstances.

The topic of the impact of technology on human socioeconomic systems is a complex one, the full treatment of which is outside the realm of the present discussion. However, it is useful to briefly describe some aspects of this important issue.

One way to illustrate the interactions of technological systems with socioeconomic systems is from a perspective of technological system requirements, applications, and impacts. Requirements are the design specifications that a new technology must achieve to successfully solve the intended problem. The issue of requirements was addressed in an earlier chapter.

Assuming the new technology successfully meets the relevant requirements, it will find application. Applications are the problems that the technological system is able solve. Some applications will be those for which the technological system was developed, however people may implement other applications once the technology exists.

Once the technology is being utilized in particular applications, physical reality is changed. A real tangible object that did not previously exist is now part of the environment. This change to humans' physical environment will have impacts. People and organizations may react to whatever changes occur. Human behaviors, ideas, assumptions may change. Human patterns of activity may be altered. Impacts then might be described by changes that occur when new technological systems enter into use. As new technology is constantly being introduced the impact process is continuous and different segments of the human socioeconomic system may experience different impacts.

## 7.7.1    Example: Incandescent Electric Light Bulb

A brief example will demonstrate some features of impacts of technological systems. Impacts are potentially more easily discerned through historical case studies. The goal is to provide an illustrative example of some highlights rather than a comprehensive analysis of effects of new technology.

The invention and adoption of the incandescent electric light bulb as a source of artificial illumination is characterized by a particularly active period from about 1878–1890.

During this time the basic electric light bulb was patented and underwent many improvements. Several hundred electricity generation stations were built. Electric lighting systems were installed in commercial buildings, factories, public spaces such as street lightings, and some private dwellings in numerous locations in the United States and European countries. With the establishment of alternating current (AC) as the standard for utility systems and with continued improvements to the bulb design and development of low-cost manufacturing methods, adoption of incandescent electric lighting grew rapidly after about 1900. By 1930 about two-thirds of American homes had electricity and electric lighting.

For electric lighting to replace candles, gas, and oil lamps in solving the problem of providing an artificial source of illumination, certain requirements had to be met. These include being sufficiently bright, safe, long lasting, affordable, and easy to use. As these requirements were fulfilled applications of electric lighting occurred in factories, stores, banks, printers, for street lighting, illuminated signs and advertising, decorations, and lighting in public buildings and homes.

The technological system of artificial illumination by incandescent electric lighting changed the physical environment. Impacts or changes in people's life patterns and behaviors occurred. Work could be done effectively at night. Night shifts in factories developed. In stores illuminated with electric light shopping could be done at night. People were able to read and work at home more easily at night. Human existence moved away from activity dictated by the need for daylight.

As the electric light become part of the human environment, building design changed to reflect less reliance on the need for natural illumination. One place where this impact can be seen is in factory buildings. Prior to electric lighting, factory buildings were often characterized by extremely large windows. This was to allow sufficient daylight into the factory so workers could see well enough to accomplish their tasks. With the introduction of electric lighting factory buildings dispensed with windows almost entirely. Figure 7.9 illustrates the contrast in factory windows before and after the availability of electric light.

## 7.7.2   Example: Transistor Amplifier

The transistor amplifier provides another brief example of the impact of a technological system on human socioeconomic systems. In the case of the transistor, the most significant impacts were not necessarily the originally envisioned applications of the technology but rather in applications different from and subsequent to the originally planned use. The transistor's biggest impacts were not from the application for which it was initially developed. This section provides a brief overview of some of the main features of a complex episode in the history of technology.

The transistor can amplify a weak electrical current. Prior to transistors, vacuum tubes dominated in carrying out this function. Figure 7.10 shows a vacuum tube and a transistor.

**(a)**                                              **(b)**

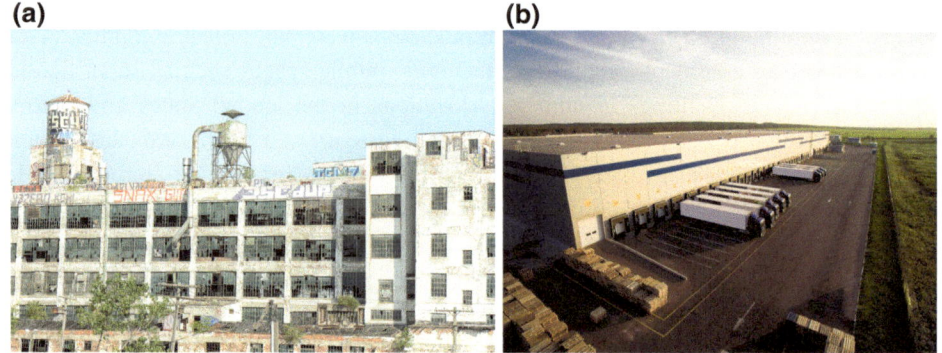

**Fig. 7.9** Contrasting use of windows in old and new factory buildings: **a** old factory (iStock.com/Gerville); **b** modern factory (iStock.com/valtron84)

**Fig. 7.10** Vacuum tube and
transistor [cropped] (iStock.
com/vlabo)

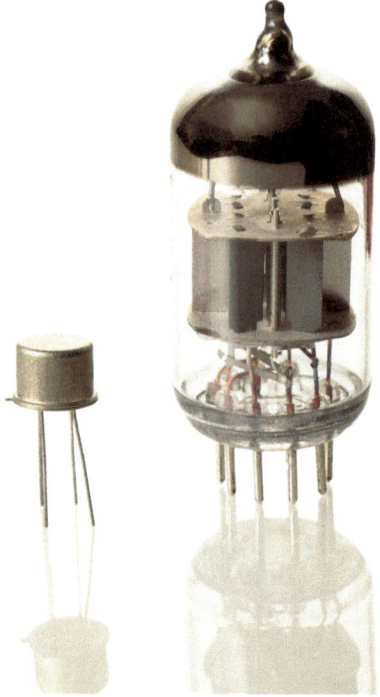

The development of the silicon transistor as a common practical device benefited significantly from work done at Bell Telephone Laboratories. A significant era of development started during World War II and extended to about 1957. Over this time the transistor went from early demonstration to mass-market implementation.

The transistor was initially part of an effort by Bell Labs to develop an improved device for radar receivers. The radar receiver had a need to "mix" or combine two input signals into a single output signal. A device based on the element germanium was thought to be a promising approach.

Gradually improved understanding of the physics of electron flow in solids resulted in development of devices based on the element silicon. The name "transistor" was coined by Bell Labs to describe a newly developed "sold-state" device that could magnify or amplify a weak signal. The "solid-state" descriptor was used to distinguish the new transistor from the already existing vacuum tube. The transistor was a small solid piece of silicon material while the vacuum tube encased the active components inside a hollow glass tube.

Development throughout the 1950s and into the 1960 s resulted in transistors that were well-suited to the requirements of consumer electronics. Transistors were small, durable, used relatively low amounts of electric power, and were long-lived. Although it was not Bell Lab's original goal, the transistor surpassed the vacuum tube in many applications. Compared to the transistor vacuum tubes were larger, required considerably more electric power, were more fragile, had shorter life spans, and generated heat during use.

One of the first mass-market applications of the transistor was in the "transistor radio" that become popular starting in the late 1950s. The small size and low power requirements of the transistor made small battery-powered radios a practical option. Prior to this time radios based on vacuum tubes were the more common technology. Due to the power requirement vacuum tube radios needed to be "plugged in" to a wall outlet. Tube portability was not a practical option for consumer vacuum-tube radios.

A significant impact of the application of the transistor to radios was the creation of personal or individualized use of electronics. Vacuum tube radios were relatively large and had to be plugged in. When used they would be heard by everyone in the vicinity. Thus, radios were a shared appliance not unlike the family stove or refrigerator. There was one in a room, used and heard by everyone simultaneously. Transistor radios created personal private use of electronics that continues to be the norm today. As transistor radios proliferated, radio stations developed greater market specialization catering broadcasts to particular audiences and segments of the population (Fig. 7.11).

The transistor also replaced vacuum tubes and found application for use in computers, although primarily as an electrically controlled switch rather than a signal amplifier. Transistors in a drastically miniaturized form as integrated circuits enabled the development of the computer, another application for which it was not initially developed. The vast socioeconomics impacts of the computer barely need mention.

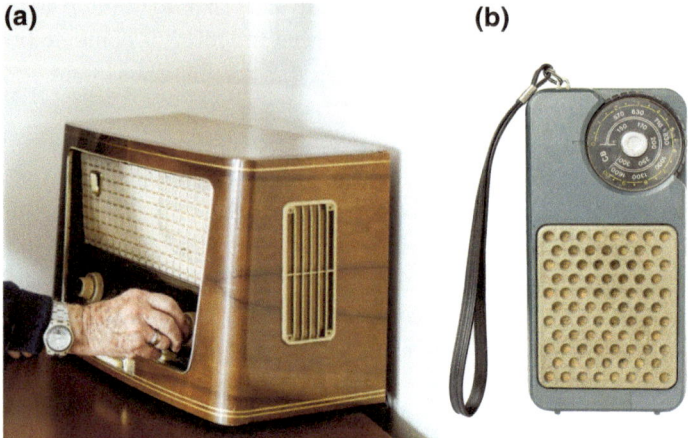

**Fig. 7.11**   Vacuum tube radio and an early transistor radio: **a** vacuum tube radio (iStock.com/ozgurc ankaya); **b** early transistor radio (iStock.com/Ballun)

## 7.8   Directing Technological Development

This section addresses aspects of the development of technological systems where the role of decisions made by people is necessary and especially evident. Analysis of the factors influencing the development of technology is a broad area that is the subject of its own discipline of study. The objective here is not a comprehensive analysis of the control of technological directions. The goal is to use the description of technology as systems of interacting components as a means of highlighting aspects of the process where human input is necessary and consequential. Identifying the locations and nature of human input from a technological systems point of view may add some helpful information to the broad questions of how the direction of technological developments are determined and influenced.

It is important to not overlook the basic fact that creating technology requires deliberate human action. Technological systems do not create themselves (yet). The raw materials in crude form, and the underlying principles of natural phenomena exist independently of people. However, technology is made through deliberate intentions of people. Different aspects of the design, production, distribution, and use involve different people and different required degrees of human agency including individuals, groups, corporations, government entities, and consumers.

Technology is created to solve problems, but the problem is identified and defined by people. Individuals or groups that are able to frame and define problems exert considerable influence over subsequent technological system development. Technological systems have a specific function expressed in verb-noun form as some kind of action on some "thing." So the fundamental purpose of technological systems is to create a change in the physical environment. Identification of the condition needing change or problem identification of a situation that can be resolved by creation of a physical object requires decision making by people.

With a problem identified, the requirements that characterize a solution are another aspect where human input influences technological developments. Requirements are established for the function and form of an acceptable solution. Function describes what the system must do, or the transformations enacted by the system. Form requirements impose conditions on form features such as cost, size, and color. Most technological systems have multiple classes of customers, users, and other stakeholders that influence requirements. These multiple stakeholders establish the requirements a system must fulfill to be allowed to operate within the broader socio-economic system and be adopted by some class of consumers.

Development, production, and distribution of technological systems requires allocation and access to resources. Resources include physical objects such as natural resources utilized by technological system construction and operation. Control of access to physical materials and natural resources involves complex interactions of individual, corporate, and collective government interests. These socio-economic mechanisms are overlapping and not necessarily sequential in influencing technological system development. In parallel with access to diverse physical resources, creation and development of technology systems requires allocation of human capital and information resources.

Besides natural resources, operation of technological systems requires integration into the sociotechnical environment. This involves many actors at multiple levels in a network of interacting systems. Interconnection to other systems such as legal, commercial, and cultural involve human decisions and deliberate actions to accomplish these integrations.

Technology depends on utilization of the phenomena of nature and involves physical construction that in some way is traceable back to raw materials from the earth. However human agency determines the function and form of technological systems starting with choosing the problems to be addressed, defining the technological solutions to these problems, selecting physical, human, and socioeconomic resources to develop the technological solutions, and optimizing conditions for integrating the new technologies into preexisting systems. Human agency in the construction of new technologies is manifested in the decisions of individuals, teams and groups within organizations, deliberate decisions of public and private entities, and purchasing decision of diverse classes of consumers. How these decisions are made involve social, cultural, economic, and governmental processes that can vary with time and circumstance.

## Bibliography

Bijker, Wiebe E., and John Law. *Shaping Technology/Building Society: Studies in Sociotechnical Change*. Brooks/Cole, 1992.

Bijker, Wiebe E., Thomas Parke Hughes, and Trevor Pinch. *The Social Construction of Technological Systems, Anniversary Edition: New Directions in the Sociology and History of Technology*. MIT Press, 2012.

Billington, David P. *The Innovators: The Engineering Pioneers Who Transformed America*. Wiley, 1996.

Billington, David P. *Power, Speed, and Form: Engineers and the Making of the Twentieth Century*. Princeton University Press, 2022.

"Electricity Generation, Capacity, and Sales in the United States - U.S. Energy Information Administration (EIA)." Accessed August 1, 2022. https://www.eia.gov/energyexplained/electricity/ele ctricity-in-the-us-generation-capacity-and-sales.php.

Ellul, Jacques. *The Technological Society*. Knopf Doubleday Publishing Group, 2021.

Groover, Mikell P. *Fundamentals of Modern Manufacturing: Materials, Processes, and Systems*. John Wiley & Sons, 2010.

Halderman, James. *Automotive Technology: Principles, Diagnosis, and Service*. Pearson/Prentice Hall, 2020.

Hughes, Thomas Parke. *Networks of Power: Electrification in Western Society, 1880–1930*. Johns Hopkins University Press, 1993.

Hughes, Thomas Parke. *Human-Built World: How to Think about Technology and Culture*. University of Chicago Press, 2004.

Mumford, Lewis. *Technics and Civilization*. University of Chicago Press, 2010.

Pool, Robert. *Beyond Engineering: How Society Shapes Technology*. Oxford University Press, 1997.

Riley, Donna. *Engineering and Social Justice*. Synthesis Lectures on Engineers, Technology, & Society. Springer International Publishing, 2008. https://doi.org/10.1007/978-3-031-79940-2.

Riordan, Michael, and Lillian Hoddeson. *Crystal Fire: The Birth of the Information Age*. W. W. Norton & Company, 1997.

# Systems-Level Similarity: Technological Domains or Clusters

**8**

## 8.1 Chapter Overview

- Technological domains are groups of technological systems that are related by use of a common core system and application of the same underlying physical principles.
- Systems within a domain address related problems by modifying a basic system through means such as changes in scale, addition of functions through specific components or subsystems, and adjustment of component forms. Industries often develop around technological domains. This includes entities that are responsible for development of major technological systems and other groups that produce components or subsystems.
- Standards of practice and legal codes may be established addressing system features and characteristics of systems within the domain.
- Engineers often work within a particular technological domain. Through training and experience they acquire expertise in utilizing the families of components and common function structures within the domain. Engineers are able to create variations of systems within the domain to meet a specific function accessible within the domain.
- Different technological domains intersect and interact in complex ways. The variability of determining the boundary between a component and a system when analyzing a particular element of technology is reflected in the technological domain as well.
- A system in one technological domain might be considered as a component in another.

J. Krupczak, Jr., *Understanding Technological Systems*, Synthesis Lectures on Engineering, Science, and Technology, https://doi.org/10.1007/978-3-031-45441-7_8

## 8.2    Technological Domains, Clusters, or Families

Engineering design domains exist around collections of related components. Over time, technological systems domains develop which are based around particular sets of components and physical principles. Other terms, such as clusters or families, could describe these groups of related systems. Systems within a domain or family share common components and core underlying principles.

Systems within a domain or family are related but they are not identical. Systems consist of both components and function structures that are characteristic of the domain as well as components unique to that particular system. A combination of the core components of the domain, variations in component scale, and system-specific adaptations make it possible to efficiently address a range of problems.

Systems within a domain may provide the same overall function in diverse applications. These domains often develop around a specific product or related applications and problems to be solved. Engineers will frequently work within a particular domain and develop extensive expertise in utilizing the families of components to create systems within the domain to meet a specific function.

## 8.3    Technological Domain Examples

Over time, technological systems develop into groups that have a similarity at the system level.

These systems have some components in common. The pattern of interactions between major components is the same for systems in the group. The same set of core underlying physical principles describes major phenomena utilized in the related systems. While the systems are similar, they are not identical. These groups might also be described as technological clusters, families, or domains.

Figure 8.1 shows a group of technological systems. They are quickly recognized as aircraft. Upon closer analysis, differences become apparent. The collection includes a modern military jet, a passenger airplane, a World War I era plane, and a toy model aircraft. All of these are immediately recognized as flying machines despite differences in size, age, and purposes. The use of a shared set of components and structures facilitates recognition of these as aircraft. For instance, each uses the characteristic airfoil wing, has similar tail section design, and the major components are organized into a similar overall structure. The same underlying physical processes describe major aspects of how these systems accomplish desired functions.

Other examples convey the idea of technological systems related by shared components, structure, and underlying principles. Figure 8.2 shows four structures: a bridge, an antenna tower, a tower to support electrical wires, and a building structure. A "family resemblance" is recognized. The four structures show a similarity of design elements.

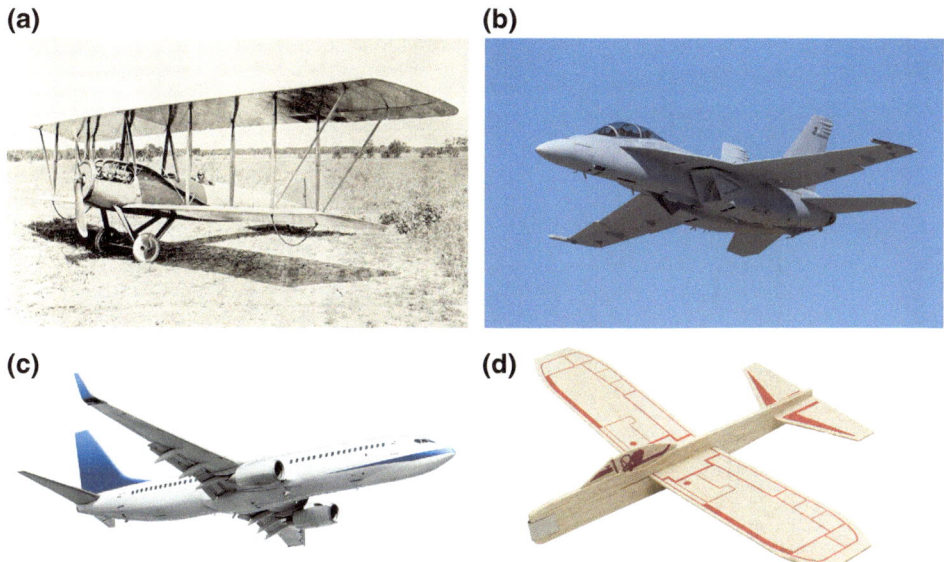

**Fig. 8.1** Several aircraft: **a** biplane (iStock.com/andipantz); **b** military jet (iStock.com/JohnnyPowell); **c** commercial airplane (iStock.com/phive2015); **d** toy airplane (iStock.com/DNY59)

Each uses some of the same components in different circumstances. An underlying similarity of structure can be seen. However, each technology is used for a different application and the arrangement and interactions of the components are not identical.

It is not difficult to find other technological systems that by casual visual inspection, seem related to others in a group through similar components, interactions, and structures. Figure 8.3 shows the internal circuitry of several electronic devices. Electronics is an example of the use of common elements in devices spanning a vast range of applications and multiple layers of hierarchy.

Another example is found in chemical processing facilities which display an underlying similarity of form. Figure 8.4 depicts a crude oil refinery, a facility making ethanol from corn, a plant to extract and separate nitrogen and oxygen from the air. Each displays common characteristics of system organization despite differing end products. The similarity of components within the chemical processing facilities indicates that some subfunctions internal to the systems are the same.

The construct of domains of similar technological systems described here is not characterized by any absolute quantitative criteria but rather as a continuum of different degrees of similarity between systems. This means of organization can be helpful in classifying the thousands of different technological systems developed by people for diverse applications. The idea is a helpful organizational framework similar to categorization and associations of living things in a diverse and complex biological ecosystem.

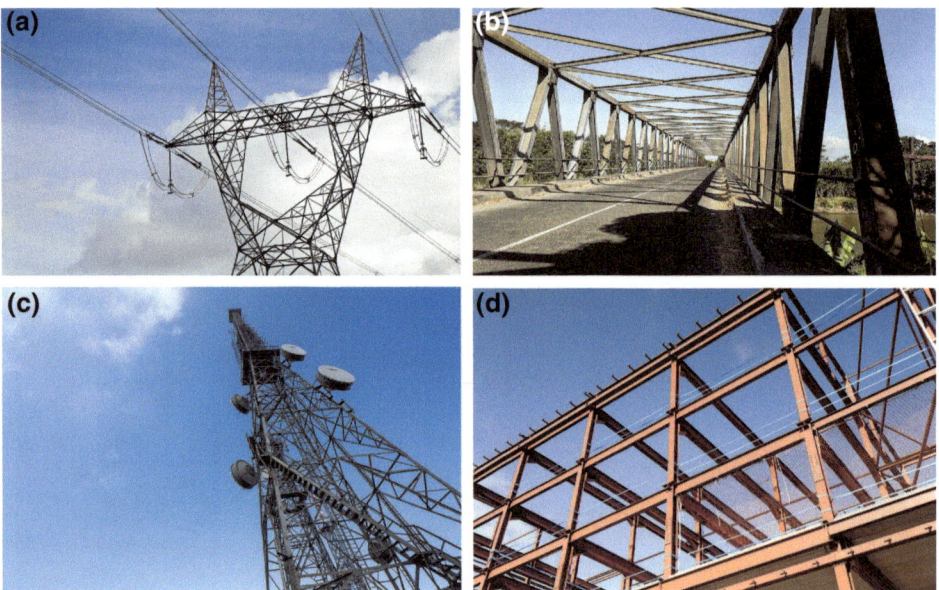

**Fig. 8.2**  Four structures: **a** powerline (iStock.com/travenian); **b** steel frame bridge (iStock.com/Rak hmat Sobirin); **c** signal tower (iStock.com/onlyyouqj); **d** steel frame building (iStock.com/Stephen Barnes)

Consider the group of technological devices shown in Fig. 8.5: a lawn mower, an ultra-light airplane, power washer, an emergency power generator. These form an example of limited similarity within the group. The surface appearance is very different. These devices have very different uses and applications. However, each uses a small internal combustion engine. The engines are nearly identical. The engines interconnect with different components in each system. The internal structure of the component interactions differs considerably between these objects. This collection is an example of a component, the small internal combustion engine, finding application in different systems. However, the systems-level similarities are limited.

Another group of technological devices is able to illustrate a different aspect of the nature of the similarities and differences between technological devices. Consider the group consisting of a bicycle, an electric vehicle automobile, a golf cart, a one-person mobility scooter, and a wheel chair being pushed as shown in Fig. 8.6. In this case there is no shared internal combustion engine between the various devices. Although all of these are transportation vehicles, there is a high degree of difference among them. The similarity or "family resemblance" is lower in this collection.

It is possible to identify common characteristics in the group shown in Fig. 8.6. Looking past superficial appearance reveals that similarities do exist in this group. One similarity is the use of wheels in the form of rubber tires of varying types. All also require

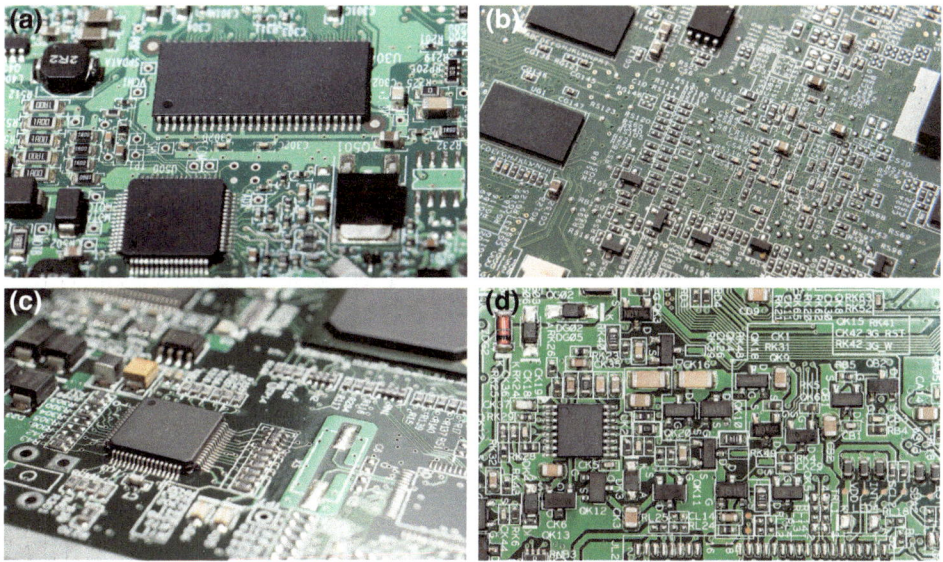

**Fig. 8.3** Internal circuitry of several electronic devices: **a** circuit board (iStock.com/eldadcarin); **b** circuit board (iStock.com/Aliaksandr Lapo); **c** circuit board (iStock.com/Tushchakorn); **d** circuit board (iStock.com/HandmadePictures)

**Fig. 8.4** Chemical processing facilities: **a** distillation columns (iStock.com/Aliaksandr Yarmash chuk); **b** oil refinery (iStock.com/Chun han); **c** bioethanol plant (iStock.com/Afransen)

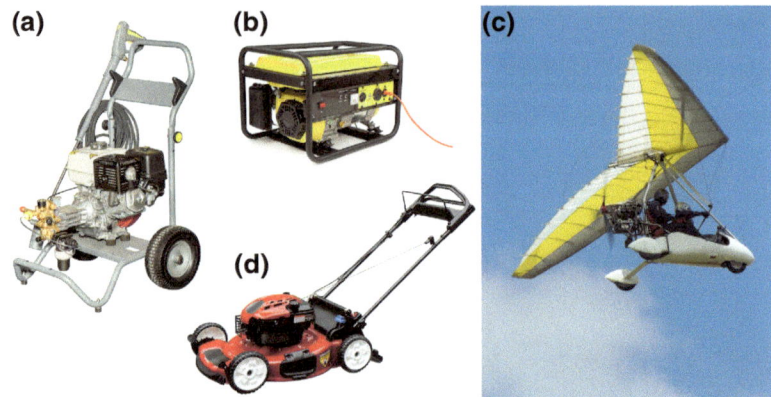

**Fig. 8.5** Internal combustion engine applications: **a** pressure washer (iStock.com/Sergii Petruk); **b** emergency electricity generator (iStock.com/DonNichols); **c** ultra-light airplane (iStock.com/aalexx); **d** lawn mower (iStock.com/ginosphotos)

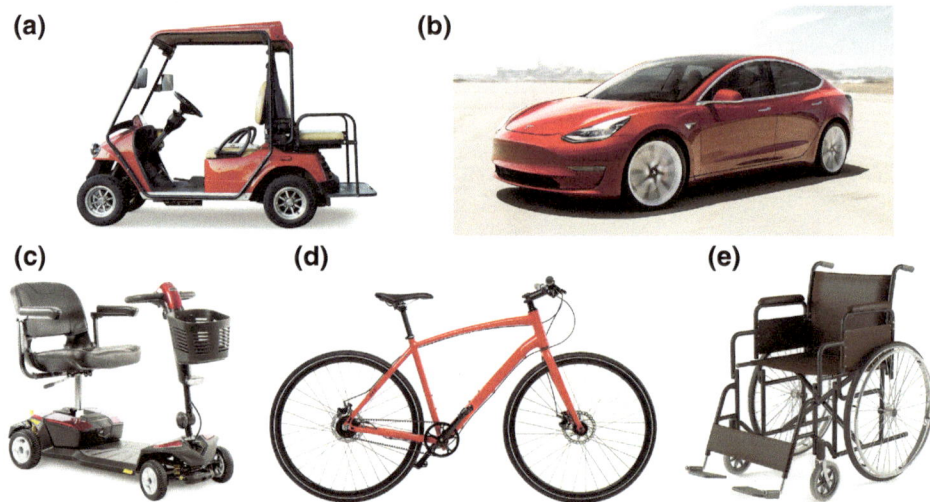

**Fig. 8.6** Examples of transportation vehicles: **a** golf-cart (iStock.com/mladn61); **b** electric automobile (iStock.com/canadianPhotographer56); **c** mobility scooter (iStock.com/CoolPhotoGirl); **d** bicycle (iStock.com/VladislavStarozhilov); **e** wheel chair (iStock.com/Flashvector)

a means of creating energy of motion. Two of these use human power, the others use different sized electric motors. All have a means of steering or controlling the direction of travel. The mechanisms differ, but the function in each case is to allow the action of the driver to impart a change of direction to the wheels. All include some type of brake system to slow and stop the vehicle using friction. All require a structure to support the forces and loads imparted to the vehicle by its contents and the circumstances of its operation. The is some general similarity between the arrangement of wheels, passenger seating, steering, and brakes. In this case system similarity can be recognized but at what might be characterized as a moderate level on a continuum from unrelated to identical.

The group of vehicles shown in Fig. 8.7 demonstrates significant system-level similarity while being far from identical. Consider a gasoline-fueled conventional passenger car, a farm tractor, military vehicle, and a large truck. Each vehicle shows a similarity of components, component interconnections, and underling principles of operation within the system. The components, however, are modified to better-suit the application requirements. Each system also includes unique components not utilized by the other systems in the group.

These vehicles all utilize an internal combustion engine. All of the engines share a similar overall architecture or design. For example, all burn fuel mixed with air. The means to supply the fuel is similar. The output of the engine is in the form of a spinning crankshaft which is turned by the action of pistons internal to the engine. At the same time, these engines display pronounced differences. The sizes are different. The fuel requirements are different. The output power varies amongst the group. These differences are related to the different purposes for which the vehicles are intended.

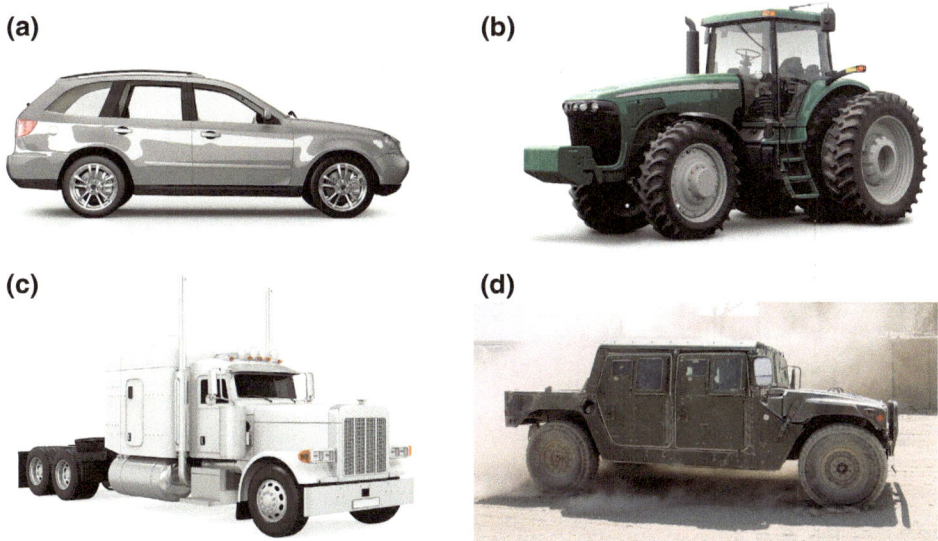

**Fig. 8.7** Examples of large special purpose vehicles: **a** passenger car (iStock.com/Henrik5000); **b** tractor (iStock.com/jondpatton); **c** semi-truck (iStock.com/Nerthuz); **d** military vehicle (iStock.com/Rockfinder)

The vehicles share similarity in other subfunctions such as wheels, brakes, steering, and accommodation of passengers. Although in each case the components of these subsystems are adjusted to the different specific requirements in the realm of transportation. Each system also includes aspects unique to its specific application. The passenger car is intended to transport one to five passengers along modern roads and highway systems. Convenience, comfort, and safety figure prominently in the requirements for accommodating the passengers. A major function of the tractor is to exert large forces on farm implements at relatively low speeds while driving in off-road conditions. The engine and wheels are adapted to this purpose. The truck system is optimized for moving heavy loads and operating at high speeds for long periods of time. The military vehicle utilizes components to function in poor road conditions, store equipment, and protect the operator.

## 8.4    Example: Home Appliances

A familiar example of a group of related systems is seen in the group of technological systems known as home appliances. Home appliances include devices like the coffeemaker, washing machines, dishwashers, blenders, toasters, vacuum cleaners, blow dryers, and some hand tools such as the electric drill. These devices are familiar due to their usefulness in carrying out specific tasks needed in everyday life. Most home appliances are based on either an electric motor, an electric heater, or both an electric heater and a motor.

Home appliances first came into existence in large numbers in the early part of the twentieth century as electric power was first being distributed to homes. Home appliances replaced human labor and heat from combustion in tasks related to the needs of daily life.

The electric motor converts electrical energy into energy of motion. In home appliances the electric motor replaced human muscles as a primer mover in implements used around the home. Appliances use electricity to generate heat and replace heat produced by combustion. Before electric heat was common, much domestic heating for tasks such as cooking was accomplished by burning wood, coal, gas, oil, or kerosene. Electric heat offers the advantages of being instantly able to turn on or off, easily controlled in intensity, and more easily contained, and safer than open flames.

Heater-based appliances include the toaster, coffee maker, hot pot, hot plate, rice cooker, space heater, toaster oven, curling iron, and clothes iron. Electric motor-based appliances include the vacuum cleaner, mixer, blender, juicer, coffee grinder, washing machine, sewing machine, electric shaver, and hand tools such as the electric drill. Some systems using both an electric motor and heater are the hairdryer, electric clothes dryer, and dishwasher. The features of the electric motor and heater are varied or optimized to meet the particular requirements of a given application. The underlying principle of operation is unchanged in each case.

In addition to an electric motor and/or an electric heater, each category of appliance includes some components that are unique to that particular device. These components provide the distinguishing functions of that appliance. A blender has a similar electric motor as a mixer, but each has components that provide functions specific to that device. The mixer attaches beater blades to the motor for mixing, while the blender attaches a rotating blade for cutting and chopping.

## 8.5   Example: Refrigeration Systems

### 8.5.1   Domestic Refrigerator

Refrigeration systems are an example of a domain of related systems. Specific systems within the domain include domestic and commercial refrigerators and freezers of various sizes and capacities, special purpose refrigerated systems like bottled drink vending machines, medical and pharmaceutical refrigerators, industrial refrigeration systems used in applications such as food processing, and refrigeration systems adapted for mobile use in transport and shipping applications. Also in this technological family are dehumidifiers, air conditioners, automotive air conditioners, and heat pumps. Figure 8.8 shows the basic vapor-compression cycle. Figure 8.9 shows some systems related by use of this cycle.

The domestic kitchen refrigerator can be used as a representative system of this domain. The primary function of the refrigerator can be described in verb-noun form as "cool-food." Other ways to express the function of the refrigerator are move–heat, or transfer–heat. These combinations emphasize that a refrigerator transfers heat from one material to another. In the case of a refrigerator the material receiving the heat is at a higher temperature than the material from which thermal energy is removed. Refrigerators transfer heat from the cool interior of the refrigerator to the warmer air in the room.

The basic principle of refrigerator operation is vaporization, a phase transition from the liquid phase to gas phase. On average, molecules in the gas phase of a substance have

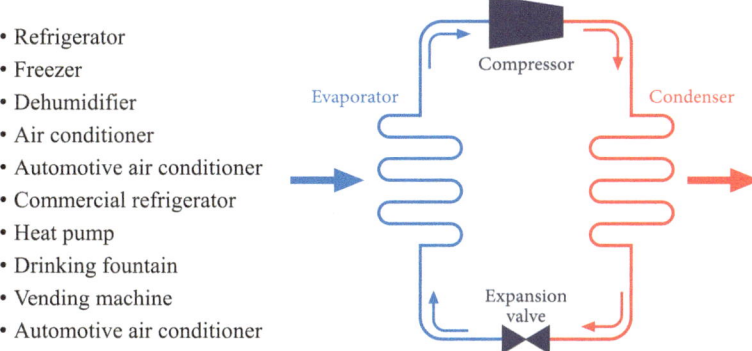

**Fig. 8.8**  Vapor-compression cycle and list of related systems (zizou7/shutterstock.com)

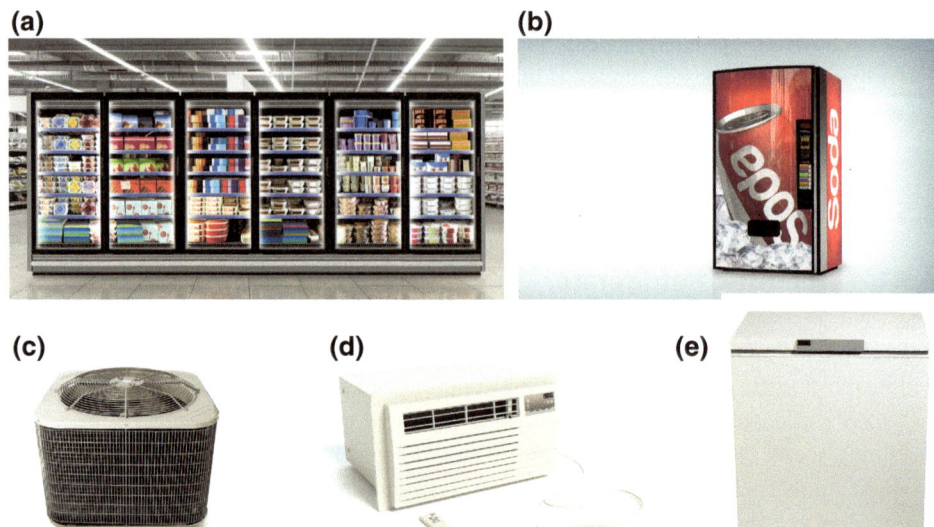

**Fig. 8.9** Related vapor-compression cycle systems: **a** supermarket freezer (iStock.com/Hitra); **b** soda vending machine (iStock.com/antorti); **c** air conditioner (iStock.com/mphillips007); **d** air conditioner (iStock.com/WesAbrams); **e** freezer (iStock.com/edoneil)

more energy than those in the liquid phase. Molecules in liquid in contact with a surface can absorb energy from the surface and in the process become a gas. The surface has lost energy in this interaction resulting in a temperature decrease of the surface.

Cooling by the transformation from liquid to gas might be familiar to anyone that has used hand sanitizer. The liquid hand sanitizer is rubbed over the hands. In the process it evaporates going from liquid to gas phase and disappearing into the air. The hands often feel cool as this evaporation takes place. Waving the hands in the air accelerates the evaporation and increases the cooling effect.

This cooling by phase change is a central effect in the refrigerator. A difference in the refrigerator is the evaporated material must be somehow captured and reused whereas the evaporating hand sanitizer is allowed to escape into the atmosphere and is not reused. The common refrigerator operates in a cycle, called the vapor-compression cycle.

The illustration in Fig. 8.10 shows the arrangement of the refrigerator in a cut-away view. Also included is a photograph of the back of a typical domestic refrigerator. The photograph shows the compressor, condenser, and expansion components.

The refrigerant liquid is confined inside a pipe or tube. The circulating liquid is typically R134-A and R410-A. These evaporate at convenient temperatures and pressures and are non-toxic and non-corrosive. In the cold part of the refrigerator, the evaporator, the liquid vaporizes. This removes heat from the inside of the refrigerator cooling the interior space and its contents.

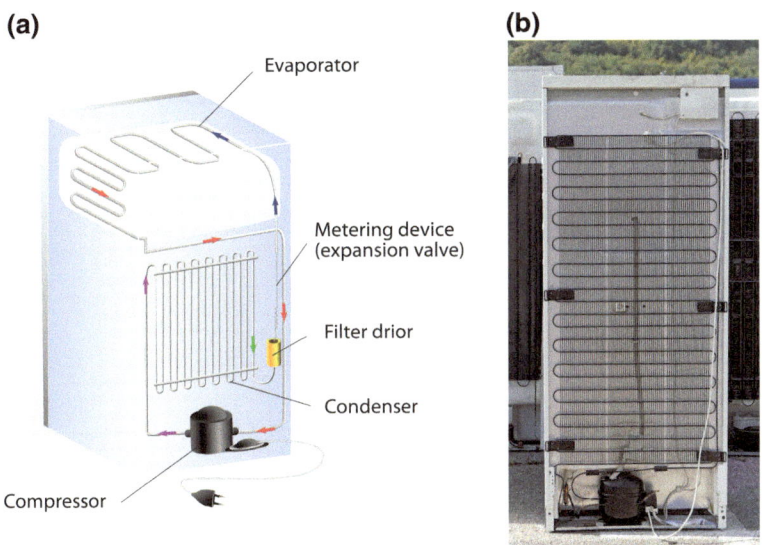

**(a)**

Evaporator

Metering device
(expansion valve)

Filter drior

Condenser

Compressor

**(b)**

**Fig. 8.10** Refrigerator system illustration and photograph: **a** refrigeration cycle diagram (iStock.
com/Designua); **b** refrigerator (iStock.com/Tatabrada)

The evaporator consists of a long tube through which the refrigerant flows. The heat
or thermal energy from the refrigerator contents has been transferred to the refrigerant by
the process of changing from liquid to gas.

The vaporized refrigerant, in the form of a gas inside the tube, must be turned back into
a liquid and reused within the refrigerator. How is this accomplished? This is a three-step
process to which the remainder of the refrigerator components are devoted. This exploits
the internal chemistry of the refrigerant material.

The vaporized refrigerant travels through the pipe to a compressor which, as the name
implies, compresses or squashes the refrigerant. This increases its pressure and density.
The temperature also increases in the process.

The compressed, warm, and high-pressure refrigerant gas must now be cooled. The
gas travels from the compressor though a pipe to the condenser. The condenser is a pipe
that loops back and forth behind or underneath the refrigerator. The warm gas in the pipe
cools transferring heat to the air in the room. The heat that was absorbed from the inside
of the refrigerator is transferred to the air in the room. While the refrigerator is cooling
food it is also warming the room.

In the condenser the refrigerant releases sufficient energy to change phase from a gas
into a liquid. It condenses from a gas back to a liquid in the condenser. At the end of this
the refrigerant is still at a high pressure but it has cooled down to near room temperature.

The expansion is the next component in the system. In this last section, the high
pressure, but room temperature, liquid must be changed into a cold liquid that can then

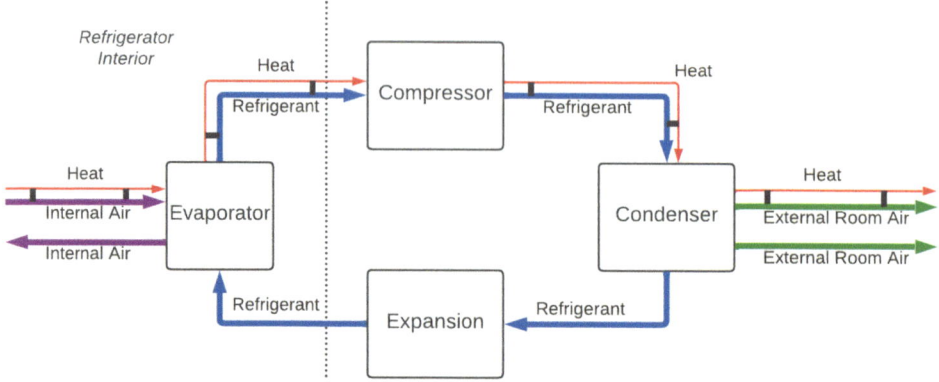

**Fig. 8.11** System diagram of a refrigerator operating with a vapor-compression cycle (illustration by author)

evaporate inside the refrigerator. The high-pressure refrigerant is sent through a valve or narrow tube into a region of low pressure. In the low-pressure region, the gas expands due to the release in pressure cooling in the process. In the home refrigerator, this expansion process to reform the cold liquid, takes place via a small tube called a capillary tube that can often be seen in the back of the refrigerator near the bottom.

The process has now gone full cycle. Cold liquid is now available to absorb heat and cool the internal contents of the refrigerator. Figure 8.11 is a refrigerator system diagram showing the main components and their interactions. The air inside the refrigerator gains heat either from conduction through the wall or from warm objects placed inside the refrigerator. The air plus heat interacts with the evaporator. The evaporator transfers the heat to the refrigerant fluid. The refrigerant fluid carries the heat around and releases it to the air in the room at the condenser located on the back of the refrigerator.

The basic vapor-compression refrigerator based on the main components of an evaporator, compressor, condenser, and expansion can be scaled in capacity for larger, smaller, and colder applications. The same basic system is used in scaled down in mini-refrigerators and scaled up to industrial freezers, refrigerated shipping containers, and walk-in coolers.

## 8.5.2  Room Air Conditioner

The air conditioner is closely related to the domestic kitchen refrigerator. Both have the same basic structure of evaporator, compressor, condenser, and expansion. However, the air conditioner provides a different function compared to the domestic refrigerator. The air conditioner cools room air while the domestic refrigerator cools the contents of an enclosed space inside the refrigerator. While the air conditioner utilizes the same structure

**(a)**                                              **(b)**

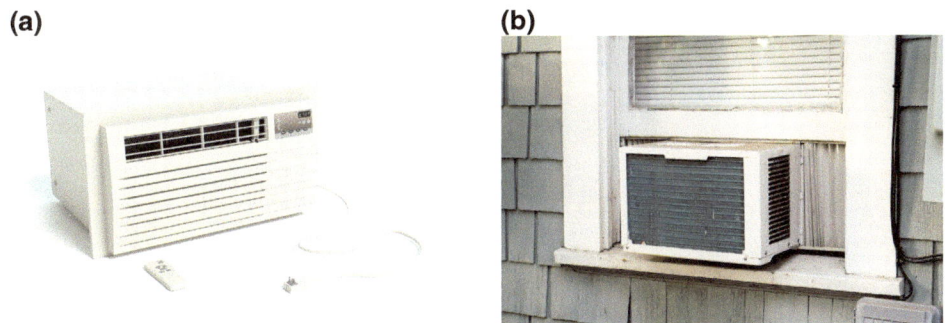

**Fig. 8.12**   Window-mounted room air conditioner: **a** (iStock.com/WesAbrams); **b** (iStock.com/Tok enPhoto)

of major components, it also has components to accomplish functions specific to cooling room air. A typical window-mounted room air conditioner is shown in Fig. 8.12.

How the air conditioner works can be illustrated by considering first a kitchen refrigerator. If the door of the refrigerator were left open, could it be used to cool down a room? Cool air would flow out of the refrigerator interior. Warm room air flowing in would be cooled by the evaporator. This process would cool the room air. However, in the rear of the refrigerator the condenser would release that heat back into the room. So, no net removal of heat would occur.

The refrigerator could cool the room if the condenser did not release the heat back into the room but rather released the heat somewhere outside the room. This is essentially how an air conditioner works.

Consider a common window air conditioner shown in Fig. 8.13. The cold evaporator is located in the room to cool the room air. The condenser is outside the building so the heat from the room is transferred to the outside air. The compressor is also usually outside since it is somewhat noisy and hot.

The air conditioner has some components that the domestic refrigerator does not have. There is a fan to move air over the evaporator. The evaporator fan promotes circulation of room air. Warm room air is drawn in and the cooled air is pushed out into the room.

There is also a fan for the condenser. This helps to circulate the outside air over the condenser to carry away the heat removed from the room. For economy a single motor is often used to turn both fans, one from either end to of the motor.

Figure 8.14 shows a system diagram for the air conditioner. A fan draws in room air containing heat. The heat is transferred to the refrigerant by the evaporator. The refrigerant carries the heat around and releases it to the outside air at the condenser. A fan helps move outside air over the condenser.

Central air conditioners are similar. The evaporator is located in an air duct. A fan circulates air through the duct to be cooled and transported to the house. The condenser

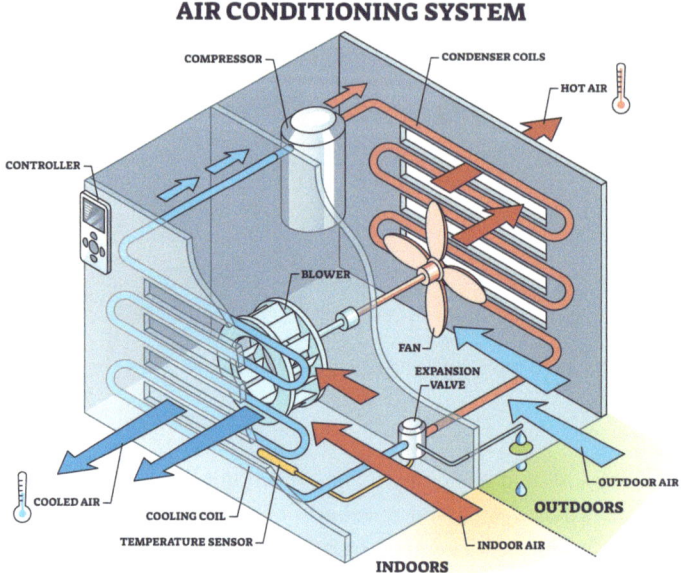

**Fig. 8.13**   Illustration of the major components of an air conditioner (VectorMine/Shutterstock.com)

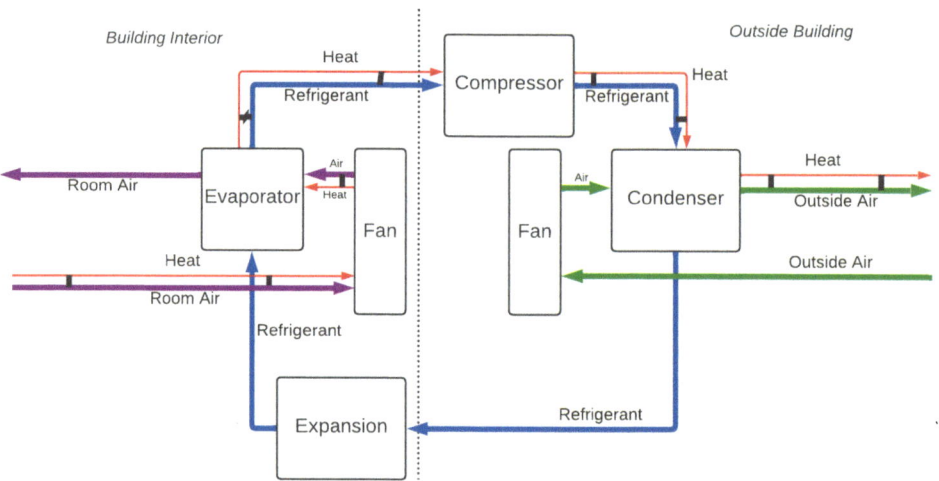

**Fig. 8.14**   Air conditioner system diagram (illustration by author)

is located outside the house. The air conditioner is seen as closely related in structure to the refrigerator. Both systems have the same main components that interact in the same way. The use of fans to promote air flow distinguishes the air conditioner system.

### 8.5.3   Dehumidifier

One common type of dehumidifier is in the refrigerator domain. Dehumidifiers remove water from air. They decrease room air humidity. Like the air conditioner the dehumidifier is based on the same vapor-compression cycle and the same fundamental components and structure. The dehumidifier also has additional components specific to the dehumidification application. A common style of dehumidifier is shown in Fig. 8.15.

The underlying principle of the dehumidifier is water vapor in the air will condense on cold surfaces. This effect is observed when cold drink bottles are exposed to warm humid air as shown in Fig. 8.15. Moisture from the air condenses into liquid on the cold bottle surface. A common dehumidifier uses this effect to remove moisture or dehumidify room air.

The dehumidifier utilizes the same main components in the same arrangement as the refrigerator. Figure 8.16 shows the arrangement of the dehumidifier evaporator, compressor, and condenser. The system is smaller and more compact but operates using a vapor-compression process.

In the dehumidifier the cold surface of the evaporator is used to condense water vapor from the air. Figure 8.17 shows a system diagram of the process. Air containing both heat and moisture enters the evaporator. The moisture condenses on the evaporator surface.

**(a)**                                                                **(b)**

**Fig. 8.15** Portable dehumidifier and water condensing on a cold surface: **a** dehumidifier (iStock. com/DonNichols); **b** condensation on plastic bottle (iStock.com/Anant_Kasetsinsombut)

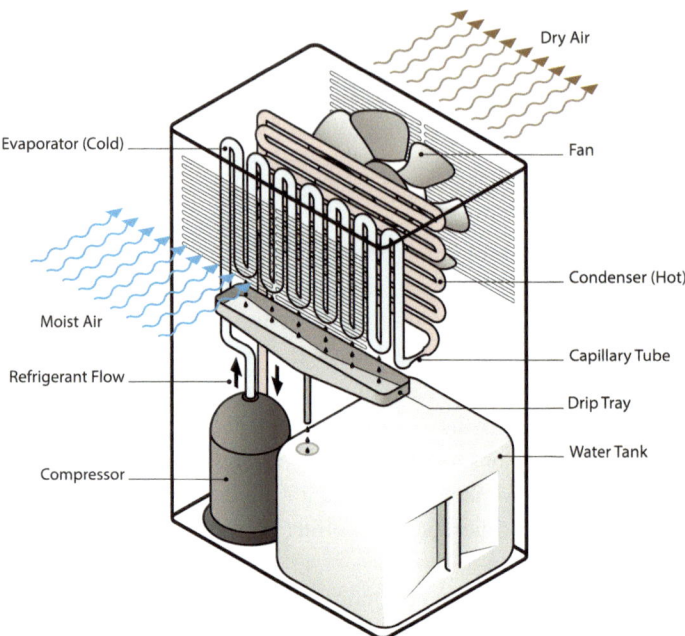

**Fig. 8.16**  Illustration of the major components of a dehumidifier (Zern Liew/Shutterstock.com)

The condensed water flows downward off the evaporator surface. Provision is made to collect the water in a tank or otherwise drain the condensed water.

In addition to losing moisture the room air is cooled when in contact with the evaporator. Heat energy is removed from the air and transferred to the refrigerant just as in

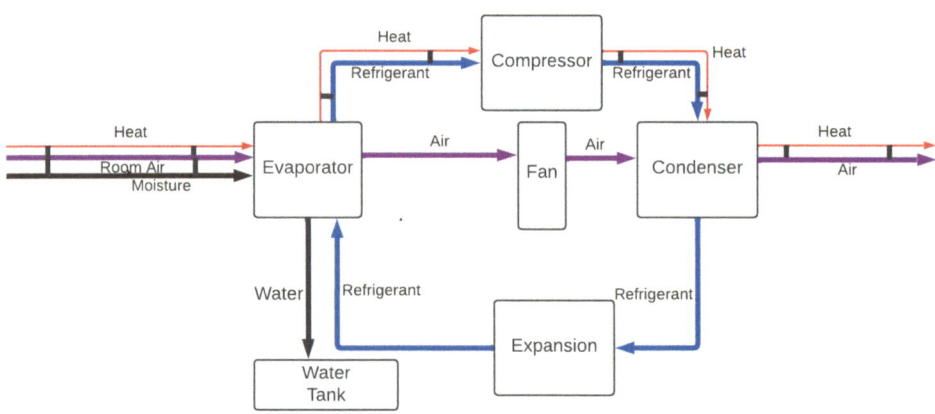

**Fig. 8.17**  Dehumidifier system diagram (illustration by author)

any other refrigerator vapor-compression cycle. Returning this air directly to the room would lead to an undesired flow of cold air. Instead, the room air leaving the evaporator is directed over the condenser. Here the heat removed from the air is returned to it. An electric heater may also be used to ensure the moisture-free air is warmed to room temperature. The air is now warmed but it does not contain the undesired humidity.

The dehumidifier uses the same core system as the refrigerator with the addition of using a fan to direct the dehumidified air over the condenser to be reheated. The humidifier also contains components to collect or drain the water removed from the air.

The examples of domestic refrigerator, air conditioner and dehumidifier show how technological systems within a domain or family share components and fundamental structures. Each system also includes some components that provide functions specific to each type of problem solved. Major underlying principles and phenomena are the same. The systems share similar mathematical models of major aspects of system behavior.

## 8.6    Codes and Standards

The existence of families of related systems leads to the issue of codes and standards. Codes and standards are established norms and legal requirements for systems and components. Codes and standards are intended to facilitate the adoption of new systems, facilitate the interconnection and interoperability of similar systems produced by different suppliers, and to protect the health and safety of the public.

### 8.6.1    Technical Standards

A technical standard is a document that defines the specific characteristics of a technological system or process. This can include dimensions, performance requirements, and safety features.

Standards establish uniform technical and engineering criteria, procedures, methods, and practices. Standards documents are usually prepared by a professional group or committee of engineers or other technical professionals with expertise in the particular type of technology for which the standard is created. Frequently professional organizations such as the Institute of Electrical and Electronics Engineers (IEEE) oversees and manages the development of standards. Standards are periodically reviewed and revised based on new developments relevant to the standard.

**Fig. 8.18** USB-C connector
cable (iStock.com/Darkwisper)

### 8.6.2  Example: USB-C Standard

A relatively simple example of a standard is the USB-C connector standard. The USB-C connector is used to transmit data and power between devices using a physical wired connection. Various digital devices utilize USB-C to interconnect with similar devices. This includes notebook computers, motherboards, tablet computers, smartphones, and hard disk drives. Figure 8.18 shows a USB-C connector.

The USB-C standard describes critical parameters such as physical dimensions of all connector features, allocation and use of each connector pin, and relevant electrical characteristics. The standard was established by the USB Promoter Group that includes representatives from Apple Inc., Hewlett-Packard Inc., Intel Corporation, Microsoft Corporation, Renesas Corporation, STMicroelectronics, and Texas Instruments.

End user consumers benefit from the USB-C standards by facilitating interconnection of components made by different manufacturers. Manufactures share the costs associated with development of this component.

### 8.6.3  Codes

Codes are similar to standards but are adopted by local, state or federal governments and can carry the force of law. A code is a set of specifications, rules, or procedures for the design, fabrication, installation, and inspection methods for a particular technological system. Codes are laws or regulations to protect public safety and health such as codes for construction of buildings. Codes protect the public by setting up minimum characteristics necessary to achieve an acceptable level of safety in products, buildings, and processes.

### 8.6.4  Example: National Electrical Code

The National Electrical Code (NEC), or NFPA 70, provides specifications for the safe installation of electrical wiring and equipment in the United States. This code was first

**Fig. 8.19** Electrical distribution panel (iStock.com/BanksPhotos)

published in 1897 and is updated every three years. The code covers definitions and rules for electrical installations including voltages, labeling, circuits and circuit protection, methods and materials for wiring such as conductors, cables, and equipment such as cords, receptacles, and switches. Also addressed are specific applications such as signs, machinery, and communication systems.

Figure 8.19 shows an electrical distribution panel similar to the type found in residences and buildings. There is a high degree of similarity between these systems in different applications but the specific implementation will be modified based on the size, use, and design of the building. The National Electrical Code helps to ensure that the design and implementation of each system avoids known problems and is within appropriate ranges of safety. Use of National Electrical Code facilitates safe and efficient design of electrical distribution systems in homes, apartments, commercial buildings, retail establishments, industrial installations, and public buildings.

### 8.6.5  Example: ASME Boiler and Pressure Vessel Code

The American Society of Mechanical Engineers (ASME) Boiler and Pressure Vessel Code is an example of an extensive code that applies to a broad class of related systems. Boilers and pressure vessels include a wide variety of fluid containment systems. Examples include pressurized storage tanks, heat exchangers, compressed air tanks, propane tanks. Also included are boilers used in industrial processing and heating systems. Figure 8.20 shows some examples of boilers and pressure vessels.

**Fig. 8.20** Examples of boilers and pressure vessels: **a** holding tanks at petrochemical plant (iStock.com/HAYKIRDI); **b** petroleum storage tanks (iStock.com/soner tuncer); **c** boiler (iStock.com/Vladdeep); **d** propane tank (iStock.com/Joe_Potato)

Boiler and pressure vessels form a group of related technological systems that are similar but not identical. They have some similar underlying functions of containing a liquid or gas under pressure that may be heated. The underlying principles influencing system behavior are similar and mathematical models for containers experiencing internal pressure are relevant to these systems.

While similar in some key functions, individual boilers and pressure vessels have substantive differences. Systems for different applications will vary in features such as size, materials contained, operating pressures and temperatures, interconnections and piping, cycles of operation, and environmental conditions in which the boiler or vessel operates. These application-specific details will influence aspects of the product form.

Safety is an extremely important concern for boilers and pressure vessels. Boilers operate at high temperatures. Pressure vessels contain gas or liquid at pressures exceeding normal atmospheric pressure. Failure of a pressure vessel can result in an explosion or the release of high-pressure gas. In industrial processes the gases or liquids contained can be corrosive or toxic. Safety issues are a strong incentive to ensure that boilers and pressure vessels function as intended for the duration of their service lifetimes.

The ASME Boiler and Pressure Vessels Code was originally established in 1914 to guide the design of boilers and pressure vessels. The code is periodically updated to include new developments and technological advances. The 2023 version spans 32 volumes.

ASME Boiler and Pressure Vessels Code describes specifications and procedures to be followed for aspects such as materials, design, fabrication, inspection and testing. The code helps design engineers to follow safe practices for features of boiler and pressure vessels and avoid known problems or unsafe conditions. The code guidelines also help to ensure the system operates as intended while making efficient use of resources.

The 32 volumes of Boiler and Pressure Vessels Code are extensive and detailed. It should be noted that the entire code does not apply to each and every boiler and pressure vessel. The code is organized to address specific components and subsystems that are relevant to some applications but not to others. Also addressed are particular special cases of interest.

Besides helping to ensure public safety, a design code such as the ASME Boiler and Pressure Vessels Code is an efficient way to preserve and transfer critical information about the design and fabrication of a particular group of technological systems. A code compiles and organizes the collective experience of the ASME in designing, constructing, and operating a group of related systems. New systems can be implemented safely and efficiently without the need to "reinvent the wheel" for this particular technology.

## 8.7   Domains and Industries

Companies and industries develop around technological domains. This includes companies that are responsible for development of major technological systems such as automotive companies and other companies that produce components or subsystems, for example, tires or windshields. The same pattern can be observed in other technological domains. In computers for example there are companies that sell what might be considered the complete system. For example companies such as Apple, HP, and Leenovo produce computers. Within this technological domain other companies produce components, subsystems, or provide services that contribute to other products within the domain. An example is Intel which makes processors.

## 8.8   Domains Intersect

Technological domains do not exist as completely self-contained entities. Domains intersect and interact in complex ways. For example, automobiles might be considered as a technological domain. An industry exists to produce a range of different types of automobiles. It forms a well-defined auto industry. However, automobiles use a considerable amount of electronics. For example, most of the operations of the engine are controlled by the ECM or electronic control module which is essentially a special purpose computer optimized for the function of controlling an automobile engine. Therefore, the automobile technological domain is seen to intersect with the microcomputer electronics domain.

## 8.9    Nature of Engineering Expertise

Engineers often work within a particular technological domain. Through training and experience they acquire expertise in utilizing the families of components within the domain. Engineers are able to create systems within the domain to meet a specific function.

Some engineering disciplines are defined around application-based domains such as: automotive engineering, aerospace engineering, and computer engineering. The subfields of the major engineering disciplines (fields defined by principal transformations such as chemical and electrical engineering) are typically application-based domains.

## 8.10    Component System Boundary and Technological Domains

The variability of determining the boundary between a component and a system when analyzing a particular element of technology is reflected in the technological domain as well. In one technological domain a particular element might be considered as a component. For example, in the domain of automobile design, the power train of the automobile might be viewed as a component. The inner workings are not considered, only the overall function of transferring power from the engine to the driven wheels. However, in another circumstance the automobile power train might be considered as a technological domain in itself. This would be the case if the power train were to be considered not as a component but rather as a system. As a system the power train is viewed as composed of individual components. An industry exists comprised of a constellation of companies that make complete power trains and power train components.

## Bibliography

American Society of Mechanical Engineers (ASME), *BPVC | Boiler and Pressure Vessel Code - ASME.*" Accessed August 1, 2023. https://www.asme.org/codes-standards/bpvc-standards.

Arthur, W. Brian. *The Nature of Technology: What It Is and How It Evolves.* Simon and Schuster, 2009.

Brookings. "America's Advanced Industries: What They Are, Where They Are, and Why They Matter." Accessed August 1, 2023. https://www.brookings.edu/articles/americas-advanced-ind ustries-what-they-are-where-they-are-and-why-they-matter/

Institute of Electrical and Electronics Engineers, "IEEE Standards." Accessed August 1, 2023. https://www.ieee.org/standards/index.html.

National Fire Protection Association "NFPA 70®: National Electrical Code®." Accessed August 2, 2023. https://www.nfpa.org/codes-and-standards/all-codes-and-standards/list-of-codes-and-standards/detail?code=70.

Reisel, John R. *Principles of Engineering Thermodynamics.* 1st edition. Boston, MA: Cengage Learning, 2015.

Stoecker, Wilbert F., and Jerold W. Jones. *Refrigeration and Air Conditioning.* McGraw-Hill, 1986.

USB Implementers Forum, "About USB-IF | USB-IF." Accessed August 1, 2023. https://www.usb.org/about.

# Technological System Evolution and Innovation

**9**

## 9.1    Chapter Overview

- The framework describing technology as systems of interacting components carrying out functions provides one perspective for characterizing the evolution of technology and the products of innovation.
- If demand and resources exist, technological systems undergo evolution and improvement to better accomplish the system function and satisfy the system requirements.
- Systems frequently undergo modification at the component level. Improvements to individual components maintain the existing system structure.
- Changes affecting multiple components frequently involve changes in the underlying principles used to carry out system functions. More fundamental changes affect more components.
- Basic structure variations tend to be most pronounced in the early stages of development after which the critical components and major principles of the system often stabilize.
- One mode of evolution involves merging of individual components into modular structures.
- In existing systems, evolution can entail a proliferation of additional capabilities and features appended to the original fundamental structure of the system, or variations of a basic system, and the addition of subsystems to extend capabilities.
- The emergence of new requirements can produce changes to systems.
- New components and materials can propagate to influence existing systems.
- Patents are granted for technological components, systems, processes, and materials, deemed novel compared to the existing state-of-the art.

© The Author(s), under exclusive license to Springer Nature Switzerland AG 2024     243
J. Krupczak, Jr., *Understanding Technological Systems*, Synthesis Lectures
on Engineering, Science, and Technology, https://doi.org/10.1007/978-3-031-45441-7_9

## 9.2    Continuous Change and Evolution

Technological systems exist in a dynamic state of ongoing change and evolution. Material resources along with economic and social conditions tend to exist in a constant state of flux. Technological systems reflect the needs and resources of the societies in which they are created. As material resources constantly change, technological systems evolve and change over time.

Analysis of the detailed mechanisms of technological change is complex and beyond the scope of this work. However, the perspective of technological systems as a network of interacting components provides one way to describe the types of changes observed as technological systems evolve. These systems transform available inputs into desired outputs while meeting functional and form-related requirements.

## 9.3    Component-Level Change

Often the gradual evolution and change in a technological system can be characterized as changes taking place at the component level of the system. In gradual and incremental evolution, the overall structure of the system is well-defined and remains constant. The function provided by the components in the system does not change. However, the performance and characteristics of the component do improve. The component may be modified and improved in some manner to enhance any desired characteristics such as cost, weight, size, durability, energy consumption, noise, or resistance to corrosion. Changes may be made to increase the performance of the component in terms of the function provided by that component in the system. The output of material, energy, or information is enhanced in some way over the previous version of the component.

An example of system change by evolution of components is shown in Fig. 9.1. The first image displays a hammer of the type used in the Stone Age. This design established the essential structure of a hammer including major components of a handle and massive head. The second image is a newer but still antique hammer. The two main components have been improved to better carry out their desired functions but the role of the components in the overall system operation is the same. The handle is no longer a stick but a piece of wood shaped to the desired profile. The stone head is replaced by a steel head that is formed to an appropriate shape. The third image shows a more modern hammer. The wooden handle is replaced by fiberglass and rubber. Across centuries, the hammer was evolved to better fulfill a desired purpose, but the basic structure remained unchanged.

Working at the component level results in a direct process for improving technological systems. The basic system structure is unchanged and therefore the function and purpose of the components are well-defined. The interactions between components are unchanged. It is possible to concentrate on finding ways in which a particular component can better meet established and well-understood requirements. The established performance and

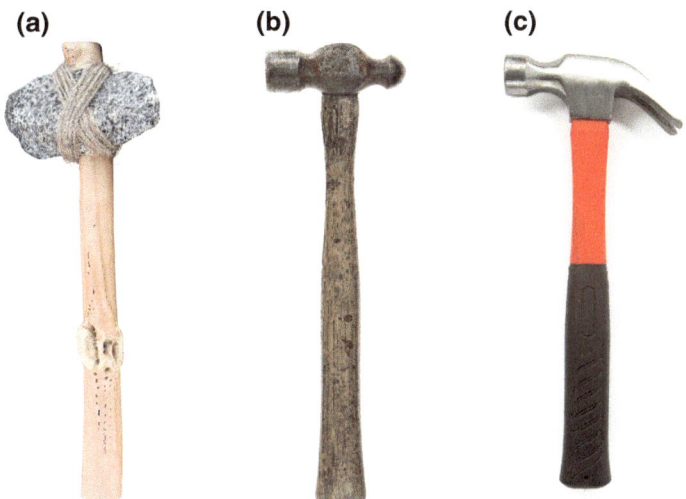

**Fig. 9.1** Evolution of the hammer: **a** stone age hammer (iStock.com/dja65); **b** wooden handled hammer with metal head (iStock.com/DebbiSmirnoff); **c** modern hammer with metal head and fiberglass handle (iStock.com/Luevanos)

operation of the system serve as a reference point from which to evaluate the benefits of any changes.

In the example of the improvements to the hammer, improvement of the handle for example, did not seek to alter the fundamental purpose of the handle in the hammer. The changes helped the handle to better carry out its established function in the system. The changes from a wooden stick up to the modern fiberglass and rubber-coated handle resulted in improvements in durability, comfort, safety, and strength within an established function of transferring force from the arm of the user to the head of the hammer.

### 9.3.1   Component Principal Change

Some technological evolution involves a change in the underlying principle of a system component. In this case, the basic structure of the system remains the same. The affected component is providing the same subfunction in the system, but the way that the subfunction is accomplished is changed.

The example of the change in aircraft propulsion from propeller to jet illustrates a change in the underlying principle of a major component. Figure 9.2 shows several aircraft starting with an early Wright brothers' model. The components providing the necessary forward thrust were a propeller driven by a piston engine. The Wright brothers' double-winged Flyer with the engine and propeller in the rear helped to establish heavier-than-air

**(a)**                                          **(b)**

Wright Flyer II (1904)                    Bleroit Monoplane (1917)
**(c)**      Top speed 10 mph          **(d)**      Top speed 47 mph

Spirit of St. Louis (1927)               P-51 Mustang (1942)
Top Speed 133 mph                    Top Speed 437 mph

**Fig. 9.2** Evolution of propeller-driven aircraft: **a** Wright Flyer II, Wright brothers, public domain, via Wikimedia Commons; **b** Bleroit Monoplane 1917 (iStock.com/guenterguni); **c** Spirit of St. Louis 1927 (iStock.com/Robert Michaud); **d** North American P-51 Mustang 1942 (iStock.com/pil esasmiles)

flight. The basic arrangement of the system changed quickly so that by about 1910 the structure of the airplane reached the now-familiar organization of a single wingspan in the middle with the engine and propeller in the front and the tail and rudder in the rear.

Over the next few decades propellers forward thrust was increased through improvements in propellers and piston engine power. This, along with other improvements such as drag reduction, was able to result in increases in top speed from 30 mph to more than 400 mph. However, the propeller-driven aircraft reached a limit at this point and further increases in top speed could not be attained.

In the World War II era the jet engine was developed. The jet engine replaced the propeller and piston engine component in the function of providing forward thrust for the airplane. While the jet engine replaced the components of the piston engine and propeller, the other components of the airplane such as the wings, control surfaces, and fuselage were unchanged. Figure 9.3 shows several early jet aircraft.

The propeller generates a force due to interaction with the air surrounding the plane. The propeller thrust is maximum when the propeller movement is perpendicular to the motion of the aircraft. As the aircraft speed increases the propeller is effectively moving at an angle through the air, following a corkscrew path rather than simply rotating in a vertical direction. Figure 9.4 shows this effect. The thrust decreases as the aircraft speed increases. Jet engines are less affected by the airspeed of the plane and jet-powered aircraft are able to obtain higher maximum speeds than those driven by propellers.

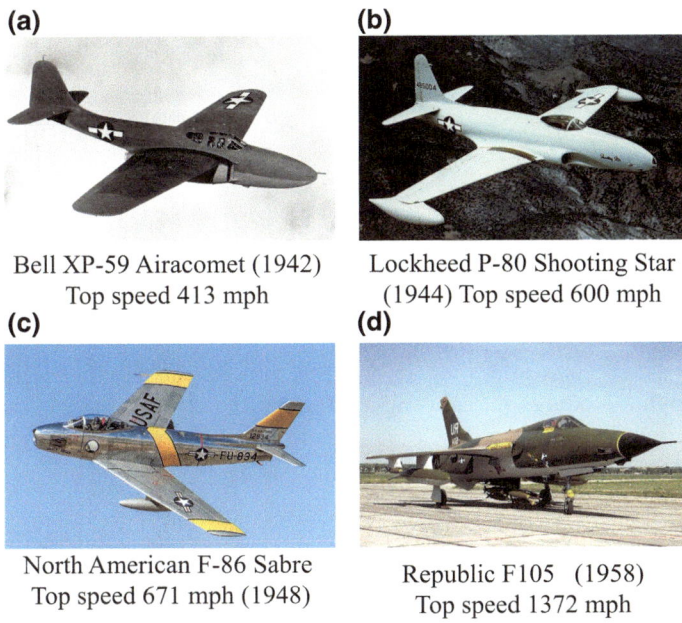

**(a)** Bell XP-59 Airacomet (1942) Top speed 413 mph

**(b)** Lockheed P-80 Shooting Star (1944) Top speed 600 mph

**(c)** North American F-86 Sabre Top speed 671 mph (1948)

**(d)** Republic F105  (1958) Top speed 1372 mph

**Fig. 9.3**  Development of jet-powered aircraft: **a** Bell XP-59 Airacomet, USAF, Public domain, via Wikimedia Commons; **b** Lockheed P-80 Shooting Star, USAF, Public domain, via Wikimedia Commons; **c** North American F-86 Sabre, USAF, Public domain, U.S. Air Force photo by J.M. Eddins Jr., Public domain, via Wikimedia Commons; **d** Republic F105, USAF, Public domain, via Wikimedia Commons

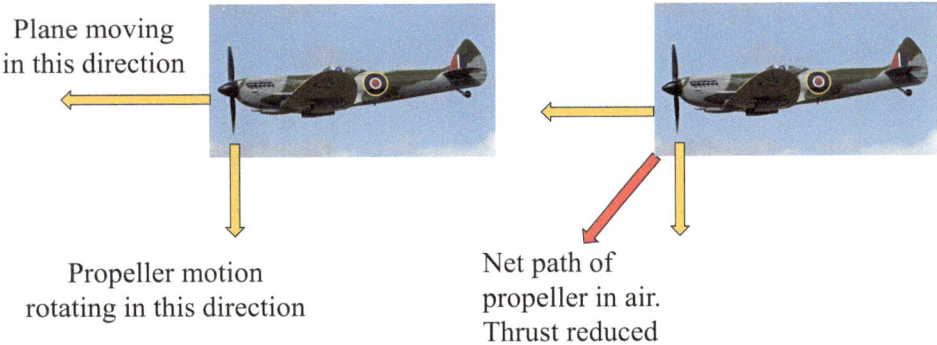

Plane moving in this direction

Propeller motion rotating in this direction

Net path of propeller in air. Thrust reduced

**Fig. 9.4**  Limits to propeller thrust: World War II Spitfire (iStock.com/serega)

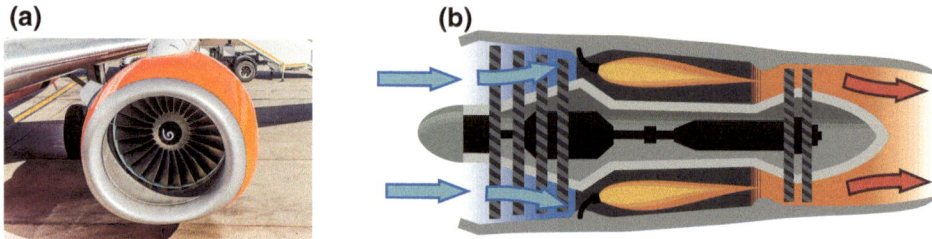

**Fig. 9.5** Jet engine thrust generation: **a** jet engine turbine (iStock.com/Vladyslav Danilin); **b** jet engine cutaway (iStock.com/jeremkim)

The jet engine produces thrust using a different underlying principle than a propeller. The jet engine burns fuel and air with the resulting exhaust ejected at high speed out the rear of the engine. This results in a force on the engine and plane in the direction opposite to the flow of the exhaust. This is illustrated in Fig. 9.5.

Aircraft powered by jet engines initially had top speeds only slightly greater than the fastest propeller-powered planes. However, the jet did not have the same limitations as the propeller and jet speeds quickly increased soon surpassing the speed of sound. The fundamental principle utilized for the generation of thrust changed from the propeller-driven to the jet-powered aircraft; however, other aspects of aircraft function such as the generation of lift by the wing surfaces and directional control remained unchanged at that time.

### 9.3.2  Multiple Component Changes

Changes can occur that encompass multiple system components leading to noticeable changes in the system structure. The overall system function remains the same. In some cases, changing multiple components addresses a specialized or refined set of requirements. The visual appearance of the system can undergo drastic alteration when several components change or are replaced.

The air wrench is an example of a change affecting multiple components in the structure of the system but not the overall function of the device. This is shown in Fig. 9.6. A socket wrench is used for tightening and loosening threaded fasteners. The arm of the user provides the force input to the system and a rachet mechanism determines whether the device is tightening or loosening.

A later development is the air wrench. In this case the force imparted to the head of the fastener is derived from the potential energy of compressed air, not the motion of a human arm. There is a pronounced change in the components and structure of the system. The air wrench is able to exert a greater force at a faster rate than achievable by human force alone. The revised wrench responds to additional system requirements

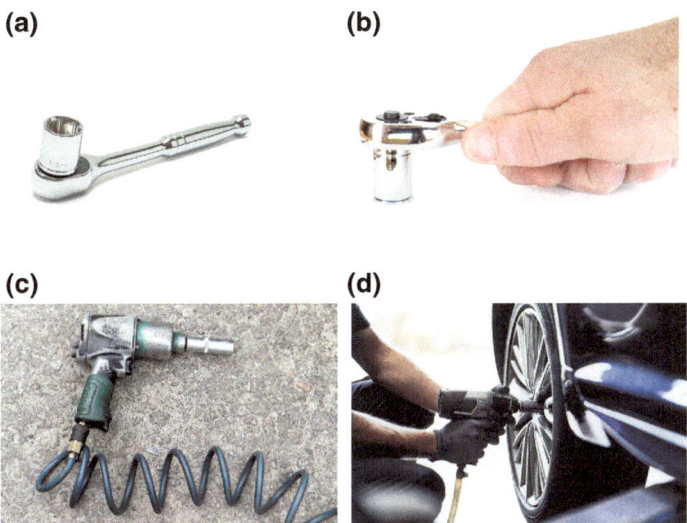

**Fig. 9.6** Socket wrench and air (pneumatic) wrench: **a** socket wrench (iStock.com/Michael Burrell); **b** ratchet and socket (iStock.com/herreid); **c** pneumatic wrench (iStock.com/Mariia Demchenko); **d** pneumatic wrench (iStock.com/BartekSzewczyk)

consistent with frequent repetitive use as in a manufacturing setting. The revised system also requires a new input in the form of compressed air.

### 9.3.3   New Components from New Science

Technological system components apply natural phenomena to achieve transformations of specific inputs into desired outputs. New understandings of the underlying principles of the natural world can result in the development of new components. These components can then find application in existing or novel technological systems.

The creation of new components illustrates an aspect of the relationship between the goals of science and the objectives of engineering. A main goal of science is to develop an understanding of the underlying principles of the natural world and to express those principles in a generalized or abstract manner. Technology is the manipulation of natural phenomenon with a particular intention or purpose. Engineering, in developing and changing technological systems, can utilize the understanding of nature accomplished in the sciences to achieve a desired function in a particular technological component or system. New understandings of the underlying principles of nature can be utilized to develop new components, materials, and processes. These new components, materials, and processes contribute to the evolution of technological systems.

**Fig. 9.7** Examples of LEDs
(Light-Emitting Diodes) (iSt
ock.com/Sandrexim)

Creating new components is one way in which engineering utilizes some of the out-comes of scientific investigations. Natural phenomena can be used to carry out desirable transformations of inputs to outputs. Components can be established to provide useful functions employing these phenomena.

The Light Emitting Diode or LED provides an example of how the understanding of natural phenomena can be used to create a component with a useful function. LEDs are based on the phenomenon of electroluminescence. In electroluminescence an electric current causes the material to emit light. The underlying effect is the recombination of negatively charged electrons with positively charged vacancies or holes and the release of energy from the electrons in this process in the form of light. A detailed understanding of the physics of semiconductors such as silicon and germanium made possible LEDs with reliable properties.

Figure 9.7 shows examples of LEDs. LEDs convert electrical energy into light. These components possess useful properties that help them to meet the system requirements of numerous consumer products. The LEDs are small in size, operate at low voltages, use relatively small amounts of electrical energy, do not give off much heat, can be fabricated to emit light of various colors, and work for very long periods of time without burning out.

### 9.3.4   New Components From New Materials

Another way in which developments in science can lead to new components is through the creation of new materials. New materials or materials with modified properties can provide forms with desirable characteristics. These new forms can be adapted to create new components and systems with capabilities not previously possible.

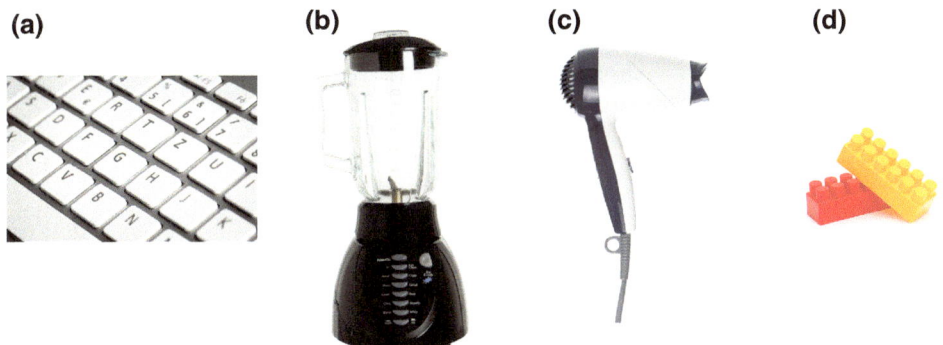

**(a)**        **(b)**        **(c)**        **(d)**

**Fig. 9.8** Familiar objects made from ABS plastic: **a** computer keyboard (iStock.com/scyther5); **b** blender (iStock.com/gregdh); **c** hair dryer (iStock.com/pepifoto); **d** Legos toy blocks (iStock.com/lena5)

Plastics are an example of new developments in science leading to new materials with useful form properties. Plastics are polymers that are created artificially through a chemical process. Plastics are not naturally occurring materials. The basic chemistry of plastics is the ability to create long chains or structures of molecules by combining smaller simpler molecules. The capacity to reliably and consistently create polymer plastics was made possible by advances in the science of chemistry and inexpensive large scale polymer production through chemical engineering.

Acrylonitrile butadiene styrene (ABS) is a polymer with many applications. ABS is formed by creating molecular chains of acrylonitrile, butadiene, and styrene. ABS can be molded at high temperatures which then becomes a rigid solid upon cooling. ABS is tough, impact resistant, and unaffected by acids, alkalis and oils. ABS can be colored using pigments.

The properties of ease of molding, impact resistance, and color variety result in ABS being used in components such as cases for consumer electronic products, the body and housings of home appliances, and elements of protective sports helmets. ABS is frequently used for the keycaps of desktop computer keyboards. Components of toys is a common ABS application. LEGO™ bricks are made of ABS. Examples of familiar objects utilizing polyethylene are shown in Fig. 9.8.

### 9.3.5  New Processes

Techniques and fabrication methods facilitate the production of components and systems. New processes for manipulating the form of materials can lead to the creation of new or improved components. Materials and their intrinsic properties are the basis of components,

**(a)**                                                                              **(b)**

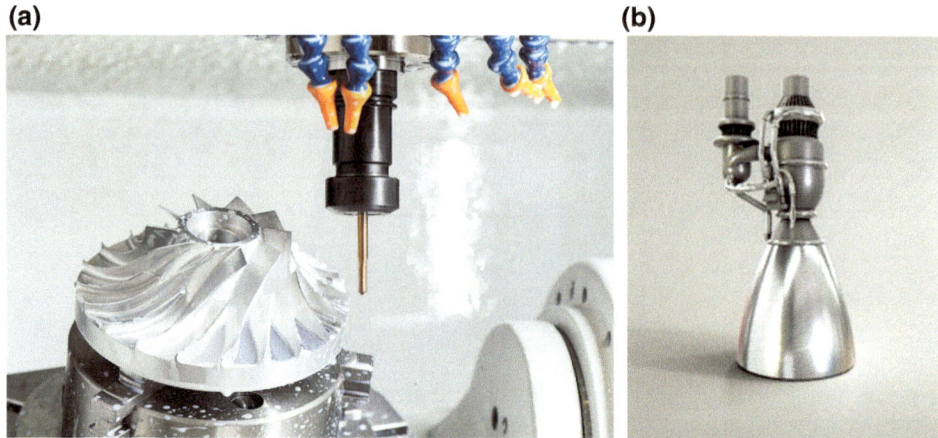

**Fig. 9.9**  Multi-axis machining: **a** 5-axis CNC milling machine (iStock.com/kadmy); **b** SpaceX Rapter Rocket Engine (Alex Terentii/Shutterstock)

however the raw materials usually require modification, particularly in geometry, to enable the component to carry out its function.

The development of multi-axis machining and computer-aided solid modeling is an example of how new fabrication techniques facilitate component creation. In multi-axis machining a cutting tool is not constrained to removing material in a straight line. Curved surfaces can be accurately cut. Using computer-aided solid modeling, the geometric shape of the component can be defined and then sent directly to the multi-axis cutting machine. In this way components with complex geometries can be designed and produced.

Figure 9.9 shows an example of a component produced using multi-axis machining. The component is an impeller similar to those used in fuel turbopumps of a rocket engine such as the SpaceX Raptor engine. The turbopumps transport the cryogenic liquid methane and liquid oxygen fuel and oxidizer. The impeller has a complex structure of interleaved curved surfaces that is cut out from a single block of metal by the multi-axis machining process.

## 9.3.6   New Analytical Methods

New analytical methods improve the ability to create new or improved components. A characteristic of modern engineering practice is the use of mathematical models of component behavior in the design of components and systems. Mathematical models can be used to predict outcomes of interest for different component parameters and operating conditions. This type of analysis helps to ensure that the components and systems will meet relevant requirements before time and resources are spent constructing and assembling the system.

A relatively new analytical technique facilitating component design is the finite element method (FEM). FEM is a numerical and computational approach for solving equations. It can be used for solving equations in mathematical models of engineering interest including stress analysis, fluid flow, heat transfer, and electromagnetics. The approach subdivides a large shape into smaller simpler shapes called finite elements. Solutions for the elements are merged into a solution for the entire geometry. Finite element analysis is computationally intensive, but its application has been made possible by steady increases in the computing power of both personal and large-scale computers. In addition, improvements in the technology of computer display screens have made it possible to easily view the calculation results.

Figure 9.10 is an example of component finite element analysis. The component shown is a bracket. FEA analysis was used to calculate the stress throughout the component under the expected use conditions. The results of the mathematical model are displayed as a color overlain on the image of the geometry of the bracket. Higher stress is indicated in red and lower stress in blue. The results can be used to determine if the expected stresses are within acceptable limits for the component material. If not, the geometry of the component can be revised and the analysis repeated until a form for the component that meets the applicable requirements is achieved.

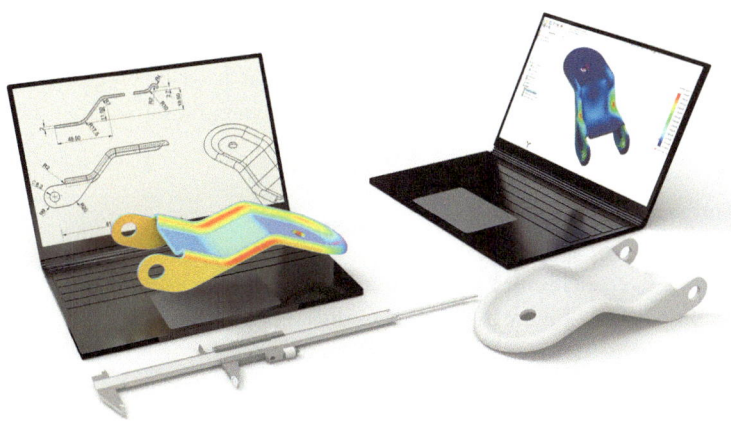

**Fig. 9.10**  Example of analysis to improve component design (3DConcepts/Shutterstock.com)

## 9.4    Separated to Integrated Components

A mode of evolution of technological systems involves transition from separated or individual components to the merging or integration of components to achieve reductions in cost and size.

Once a particular technological system has become established, one occurrence is the evolution from individual separated components to components that are integrated and merged. Separate components are fused into assemblies in which the individual components are no longer accessible. The functions carried out by individual components are combined together to create a more complex component that fulfills multiple functions in a way that is highly optimized for the particular system.

Transition from modular to integrated components occurs because integration often results in reductions of cost and size, both of which are usually beneficial improvements of technological systems. The integration of components can also lead to optimization of performance and improvement of characteristics that are useful for that system. Integration tends to increase system reliability while at the same time decreases the ability to repair or replace individual components.

Component integration is an indicator of a system that is moving into wider adoption and use. When a technological system is new, the system is undergoing initial development and the frequent revision of the function structure favors the use of individual components. In the early stages of development, a system often utilizes components initially developed for other systems which, while adequate, are not optimal for the needs of another system. When the essential operating parameters of a system have been established, movement toward integration promotes quality control and reduction in manufacturing costs. Integration or creating of specialized components is an expensive process but when demand for a system is well-established and system requirements are better understood, the overall improvements resulting from integration warrant these expenses.

### 9.4.1    Example: Microwave Oven

The microwave oven demonstrates evolution through increased specialization and integration of components. Figure 9.11 depicts different stages of microwave development. The first microwave ovens, developed by Raytheon were over six feet in height and weighted close to 800 pounds. These units employed microwave generators originally developed from Raytheon's aircraft radar surveillance systems. The components of the system were developed for applications involving installation in commercial airports and military applications.

Restaurants tended to be the first adopters of microwave ovens. Once these first models established the benefits of microwave cooking, components were modified to better suit the conditions in a domestic kitchen. The first models finding widespread adoption in the

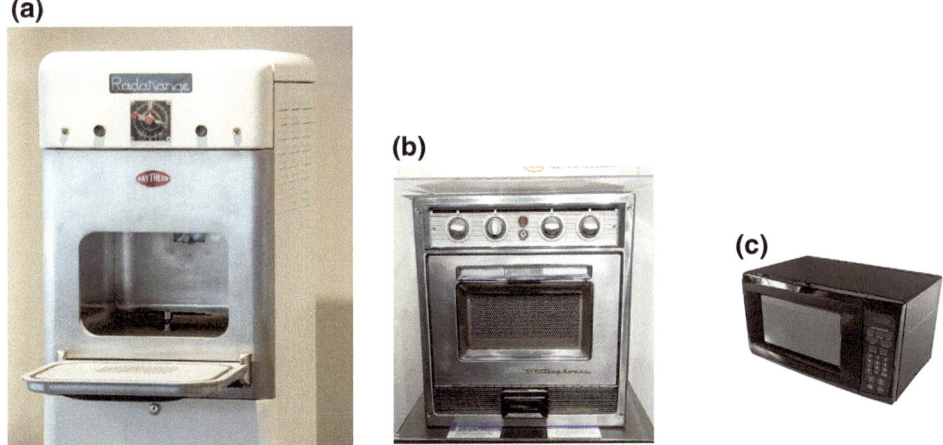

**Fig. 9.11** Reduction in microwave oven size: **a** Raytheon Radarange microwave oven, 1940's (Ray theon.com); **b** Westinghouse Microwave Oven, 1956 (Daderot, CC0, via Wikimedia Commons); **c** typical modern microwave (iStock.com/trekandshoot)

home were small enough to fit on a counter top and weighted only one-tenth that of the original models. Current microwave ovens are even lighter at only about forty pounds. As component specialization and integration reduced the overall size and weight of this device, the cooking area has stayed about the same volume.

## 9.4.2   Example: Integrated Circuit

The integrated electronic circuit is one of the classic examples of evolution from separated to combined components. Modern electronic circuits are made from combinations of a few types of basic components. These basic components include conductors, resistors, capac itors, inductors, diodes, and transistors. Prior to the development of integrated circuits, more complex electronic systems were made by interconnecting individual components. Figure 9.12a shows a typical system made with individual components.

The integrated circuit was created in part to reduce the size of a group of components. It was recognized that a limitation to reducing the size of a system was the space devoted to interconnecting the individual components. Since some of the individual components such as resistors, diodes, and transistor could all be made from silicon, it was possible to combine or integrate these individual components on the same piece of silicon material. The integration eliminated the space needed for separate packaging and interconnection. An integrated circuit component is shown in Fig. 9.12b. The individual components are too small to be recognized. The integrated circuit better fulfills the system requirement of small size.

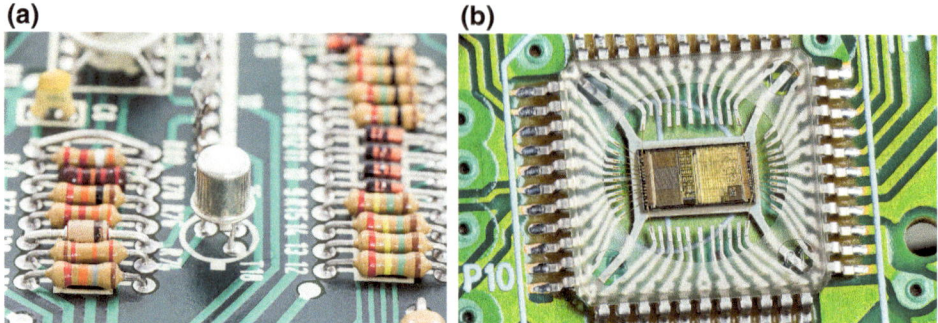

**Fig. 9.12**  Systems comprised of individual components and integrated circuits: **a** circuit board with individual components (iStock.com/stoonn); **b** close-up of electronic integrated circuit (KPixMi ning/Shutterstock)

## 9.5    System Variations in Early Stages of Development

Technological systems tend to exhibit more noticeable fundamental variations in their components and structure in the early stages of development and introduction to use. This behavior is entirely to be expected but it is interesting to observe in retrospect. The major components used to achieve important subfunctions tend to undergo the most significant changes when a system is first created. A factor that might drive this is a better understanding of system form and functional requirements as operational experience is gained along with improvements to components to better meet requirements. The allocation of subfunctions to different components is most fluid in the early stages of technological system development.

### 9.5.1    Example: The Bicycle

Examples of early versions of wheeled human-powered vehicles (later called the bicycle) are shown in Fig. 9.13. Notable variations existed in the size and configuration of wheels and the method of applying mechanical power to the wheels. The high wheeler had a very large front wheel with pedals attached directly to the wheel. In the Otto bicycle, the rider sat between two equally sized wheels. Arm and legs both applied power using a chain drive. The penny farthing bicycle positioned the rider upright between to medium-sized wheels with a third smaller trailing wheel in back for balance and steering. Pedals placed at the rider's feet transferred power to the wheels using a chain drive. Eventually the now-familiar, two-wheeled, diamond-frame design with the chain driven rear wheel became the favored configuration.

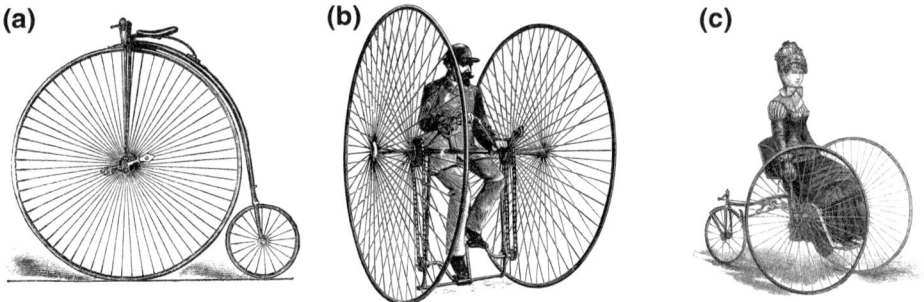

**Fig. 9.13** Examples of early versions of bicycles: **a** high wheel penny farthing bicycle from 1886 (iStock.com/Grafissimo); **b** Otto bicycle (iStock.com/NSA Digital Archive); **c** three wheeled penny farthing bike 1885 (iStock.com/Grafissimo)

### 9.5.2  Example: Digital Camera

Digital cameras are another more contemporary example. Samples of early digital cameras are included in Fig. 9.14. After an initial period of flux and change the general structure tends to stabilize. An overall optimum within the given system function and design parameters emerges. This general structure and design then remain intact as the system evolves improvements to specific components and subsystems within this general structure. One of the first digital consumer-market cameras made by Kodak was a large rectangular box shape with a hand grip on the side. A slightly later version made popular by Sony was a smaller box shape turned on the side. This model incorporated a floppy disk drive directly into the camera for storing the photos. The digital camera then became smaller and many occupied a configuration nearly identical to film cameras. Many manufacturers next migrated to smaller "pocket-size" arrangements with a viewing screen on the back. Further miniaturization resulted in the incorporation of the digital camera components into mobile phones.

## 9.6    Proliferation of System Features

In established technological systems, evolution might include a proliferation of additional capabilities and features appended to the fundamental structure of the system. After initial introduction and adoption, the systems often undergo a period of improvement which includes the addition of features that are initially secondary to the fundamental principles of operation. In this situation the fundamental operating principles of the system remain fixed. New capabilities and improvements are added but these are often not directly related to the essential operating principles and original function of the system.

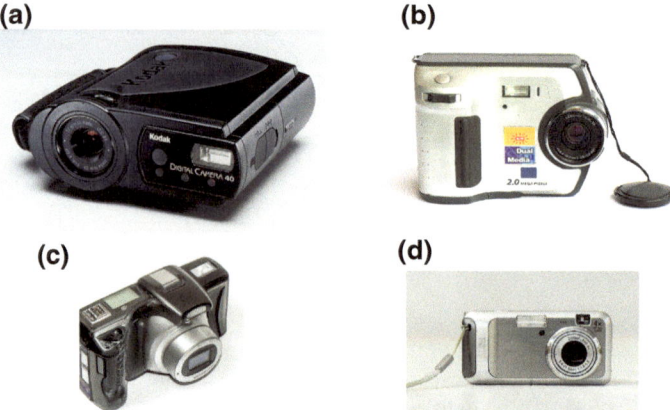

**Fig. 9.14** Evolution of the digital camera: **a** Kodak digital camera (Kodak.com); **b** early digital camera (iStock.com/singkamc); **c** early digital camera (image by author); **d** modern digital camera (iStock.com/Oleksandr Shatyrov)

The evolution of the cellphone provides an example of the accumulation of features and capabilities appended onto the basic function of the system. Figure 9.15 shows the development of the cellphone over about a 25-year period. The basic mobile phone was developed for the function of placing phone calls. As the device has evolved, the function of phone calls has remained unchanged. The mobile phone has added capabilities that exist in parallel with the function of voice communication but do not improve or change the voice-communication function. For example, a smart phone now provides additional capabilities such as: voicemail, texting, internet access, a camera and video recorder, GPS navigation, and ability to run useful applications program (apps). All of these capabilities were added to the phone to enhance the usefulness of the device for the user but these capabilities did not change the basic phone function of voice communication.

## 9.7    Change Due to Emergence of New Requirements

A mode of change for established technological systems is the addition of new requirements Technological systems that are well-established may undergo modification and change due a change in the nature of the requirements or expectations of system characteristics. The emergence of new requirements necessitates redesign of components or subsystems. New expectations or requirements frequently result in the addition of components to the system.

One example of new requirements for existing systems is new laws or other government actions that can lead to changes in technological systems. Often this takes the form

**Fig. 9.15** Evolution of the cell phone (iStock.com/yktr)

of new requirements concerning safety or environmental impacts. Existing systems may add or modify components to meet these new requirements.

For example, in the automobile the need to reduce emissions led to the addition of the catalytic converter. The catalytic converter meets a new requirement that was not previously in existence. The catalytic converter was a new component added to the automobile exhaust system to help convert some pollutants in the exhaust into less harmful gases. Similarly, coal-burning electricity generating facilities used to release coal ash into the atmosphere as part of the combustion products exiting the plant. To reduce these pollutants electrostatic precipitators were added to the exhaust stacks to remove this ash.

Another example of a change due to a new safety requirement is the use of GFCI devices on home appliances. The GFCI or Ground Fault Circuit Interrupter is used to reduce the risk of shock from an appliance. The GFCI detects an abnormal flow of current in the device and stops the flow of electrical energy to the device if this occurs. A common example of the use of GFCIs is on hair dryers. Hair dryers are often used in bathrooms around water thus increasing the risk of a shock to the user. The cord of hair dryers now includes a GFCI to reduce the possibility of an electrical shock. Figure 9.16 shows this new component added to the power cord for the device.

## 9.8  System-Level Changes

When viewed from the broadest level, many technological systems exist to provide basic human needs. Improvements address the same function but in a more desirable or improved manner. Many new technological systems are replacing something that already exists. Often the new system is utilizing a substantially changed fundamental principle of operation. This core principle change then results in a very different system with a substantially different set of interacting components than the predecessor system.

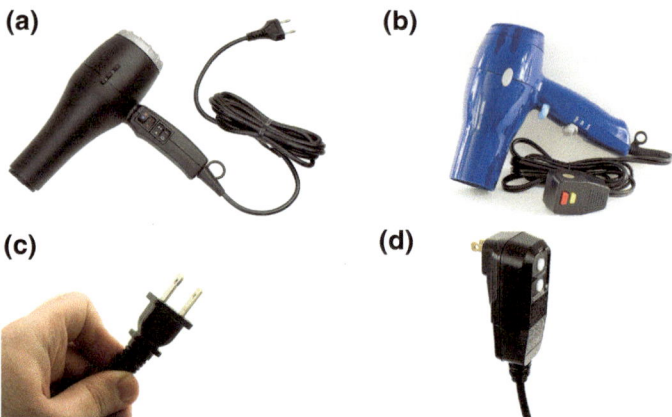

**Fig. 9.16** Example of Ground Fault Circuit Interrupter (GFCI) safety device on hair dryer power cord: **a** hair dryer with two pin plug (iStock.com/AlbertSmirnov); **b** hair dryer with GFCI plug (iStock.com/EuToch); **c** two-pin plug (iStock.com/Somus); **d** GFCI plug (iStock.com/DonNichols)

As an example, consider systems to play recorded music or in the most general terms recorded sound. Figure 9.17 shows some major stages in the development of consumer systems for listening to audio recordings. The overall function of each of these systems is to convert an input of encoded audio information into an output of audible sound. All of these systems have the same primary function.

The first system capable of reproducing sound shown in the figure is the Edison phonograph and later the gramophone record. In the recording process, air pressure variations corresponding to sound are converted into a continuous spiral groove in a cylinder or disk in which the shape of the groove corresponds to the sound pressure variations. For playback, a stylus is moved through the groove. The groove vibrates the stylus which in turn moves a diaphragm reproducing the sound. The system is entirely mechanical there are no electronic components.

A major change in the system occurred with the vinyl record player. The sound is still recorded in the form of a groove representing the sound wave but the principle used in playback has changed. A stylus is moved through the groove, but the stylus moves a coil in a magnetic field. This induces a weak electric voltage in the coil that is proportional to the original sound wave. The weak current is amplified electronically and converted back into sound in a loudspeaker using the principle of electromagnetism. One improvement in this approach over the preceding system is electronic amplification results in a much louder and clearer output sound.

A next development in mass-market consumer-oriented audio is the audio tape cassette. In this system the principle used to record the sound information is magnetism. The tape is a magnetic material, and the sound information takes the form of the level of

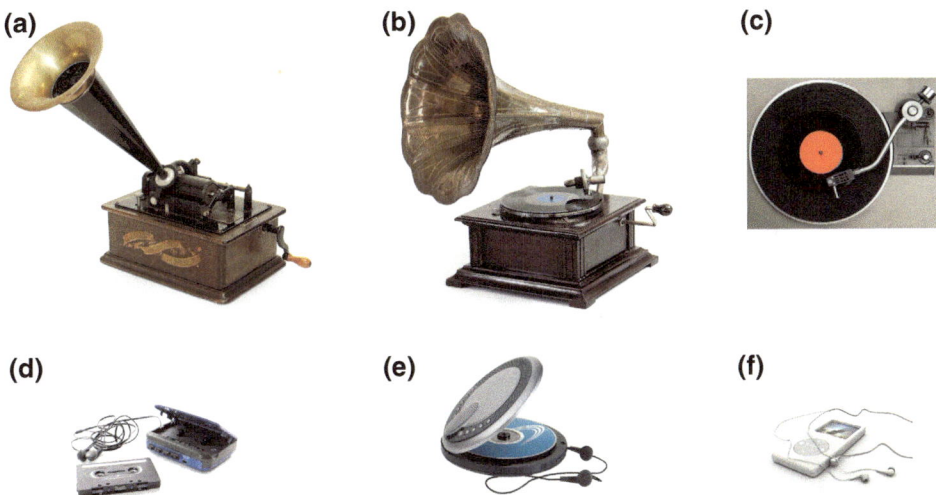

**Fig. 9.17** Examples of the evolution in audio recordings: **a** Edison phonograph with cylinder (iStock.com/igroup); **b** gramophone (iStock.com/dpullman); **c** record player with vinyl (iStock.com/kldy); **d** cassette player (iStock.com/axelbueckert); **e** audio compact disc (iStock.com/amphotora); **f** Mp3 Player (iStock.com/IvanWuPl)

magnetization at each point on the tape. For playback the principle used is electromagnetic induction. The moving magnetic tape induces an electric voltage in a coil in proportion to the magnetization level therefore creating an electric current proportional to the original sound. This is amplified and sent to a loudspeaker for conversion back into audible sound. A feature of tape cassettes is smaller size and relative insensitivity to vibration which made use while in motion possible, as in a car.

The digital audio compact disc (CD) changed the principle used to record the sound information. The original sound wave is converted into a sequence of numbers. The numbers are stored in binary 0 and 1 format in the form of small holes and flat spots called "pits" and "land" on a disk coated with aluminum. The process of reproducing the sound uses laser light. In playback a laser shines on the disk. The pattern of laser reflections of the laser back from the disk caused by the "pits" and "land" is used to reproduce the sound signal. The digital audio compact disc converted sound information into audible sound with improved sound quality and the ability to advance relatively quickly to desired locations in the recording.

The most recently developed system to convert stored sound information into audible sound is the digital audio player (DAP). Similarly to the CD, the sound signal is a sequence of numbers in binary 0 and 1 format. Rather than using a disk the binary numbers are stored in semiconductor memory inside the device. For playback the signal is read from the memory, converted back to audible sound signal, amplified, and output frequently through small speakers in headphones. The capabilities of the digital audio player compared to previous systems include, smaller size, the ability to hold substantially more audio information, the ability to be incorporated into other systems such as a mobile phone, and the capability of retrieving audio information files from external sources such as computer networks.

Consumer audio playback systems provide an example of technological system change while providing the same overall function. Each new system employed different fundamental processes for storage of the audio information and reproducing audible sound from this information. While providing the same overall function, each new system offered notable improvements in major functions and better fulfilled user requirements. Each system utilizes substantially different components. All of these improved systems, however, are still addressing the basic need of reproducing sound or audio information.

## 9.9    Patents

Patents are often associated with technological evolution and innovation. While not all technological changes are represented by patents, patents do represent some of the different modes of change observed in technological systems.

A patent is right of property granted to an inventor by a government to "exclude others from making, using, offering for sale, or selling" an invention. The patent is applicable only within the jurisdiction of that government. For example, United States patents are enforceable only within the U.S. and its territories. Patents are intended to help protect the rights of the inventor, for a limited period of time to stimulate invention and technological advancement.

In the United States utility patents are granted for the invention or discovery of a new or useful machine, process, "article of manufacture," or composition of matter as well as improvement that may be new and useful to existing items in these categories. To qualify for a patent an invention must be useful, novel, and non-obvious in relation to the existing state-of-the-art.

### 9.9.1    Example: Large Strike Face Hammer

Some example patents document the types of changes as technological components and systems evolve over time. Figure 9.18 shows U.S. Patent 8,047,099 Large Strike Face Hammer. This patent claims improvements in the head component of the hammer. The design of the strike face, or the portion of the head that strikes the object being hit, affords a larger ratio of face diameter to head size than previous designs while minimizing additional weight. Presumably the improvements to this component make it easier for the user to hit the intended object.

**Fig. 9.18** U.S. Patent
8,047,099 large strike face
hammer. Public domain

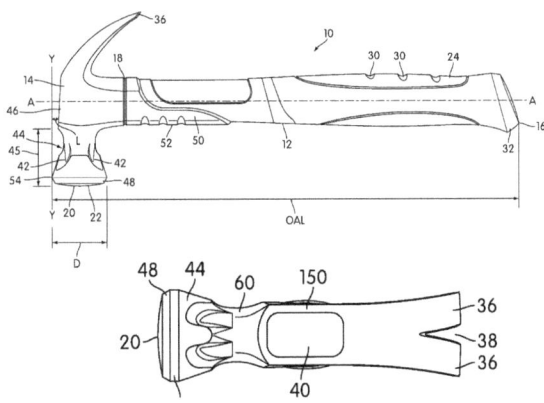

## 9.9.2 Example: The Jet Engine

Figure 9.19 shows a diagram from U.S. Patent 2,404,334 for the jet engine awarded to
Frank Whittle. The jet engine replaced the piston engine and propeller as the component
providing forward thrust in some aircraft. As is often the case, the original design of the
jet engine has undergone refinements from its original form to the design in use today.

**Fig. 9.19** U.S. Patent
2,404,334 for the jet engine.
Public domain

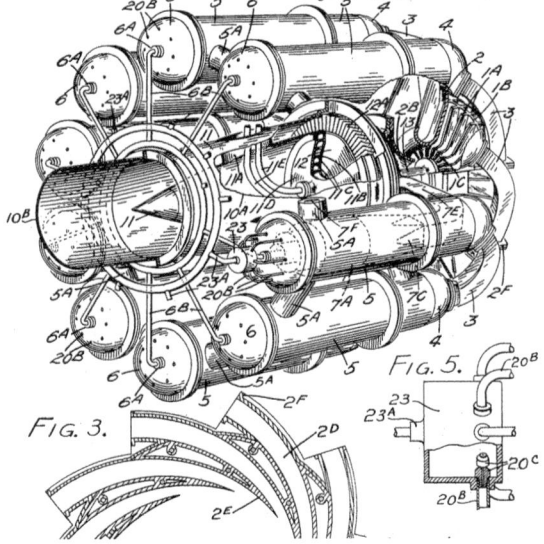

### 9.9.3    Example: The Light Emitting Diode

New scientific understandings of the underlying principles of the natural world can lead to technological components utilizing newly understood phenomenon. An earlier section described how an improved understanding of the physics of semiconductors and electroluminescence resulted in the development of the light emitting diode (LED). Figure 9.20 shows a U.S. Patent 3,293,513 for a Semiconductor Radiant Diode. This patent diagram is emphasizing the configuration of the internal semiconductor material and is not showing the cover and packaging.

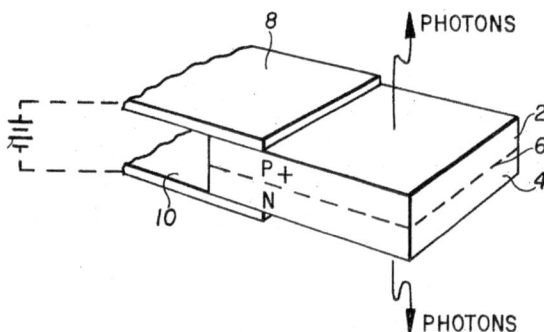

**Fig. 9.20**  U.S. Patent 3,293,513 for a semiconductor radiant diode. Public domain

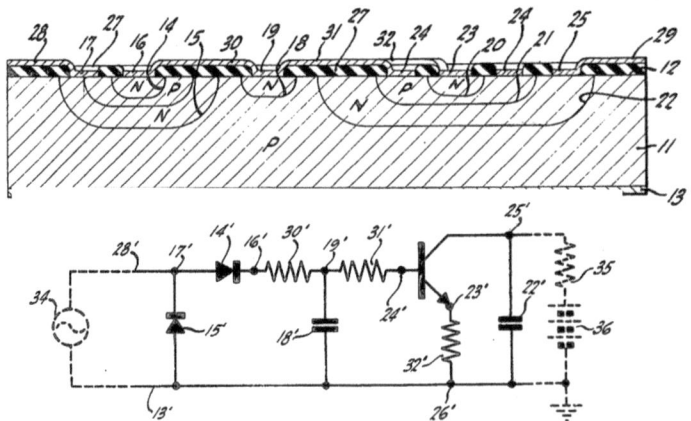

**Fig. 9.21**  U.S. Patent 2,981,877 for a semiconductor device and lead structure. Later called the integrated circuit. Public domain

### 9.9.4    Example: The Integrated Circuit

Patents can reflect the mode of technological system change in which separate components are merged into integrated units. The development of the integrated circuit was noted as an example in which individual circuit components are combined and fabricated as a single unit. Figure 9.21 is a diagram from one of the first patents for an integrated circuit design awarded to Robert. N. Noyce. The top portion of the patent diagram shows a single piece of semiconductor into which multiple components have been fabricated. The bottom portion shows a circuit of individual components that have been merged into a new single component.

### 9.9.5    Example: Kevlar

Materials are the building blocks of components, and the form properties of new materials can make possible new components and systems. Kevlar is an example of a new material that was patented under the name Aromatic Carbocyclic Polycarbonamide Fiber to inventor Stephanie Kwoleck. Kevlar is a lightweight and strong fiber with a high strength-to-weight ratio and low thermal conductivity. The exceptional strength and flexibility of

**Fig. 9.22** Kevlar: **a** police officer with a Kevlar bullet-proof vest (iStock.com/kali9); **b** excerpt from U.S. Patent 3,819,587 for Aromatic Carbocyclic Polycarbonamide Fiber, later named Kevlar. Public domain

Kevlar find use in ballistic protection equipment such as bullet proof vests, and combat helmets. Kevlar is also used in protective clothing, gloves, sports equipment, and aircraft structures (Fig. 9.22).

### 9.9.6  Example: The Quadcopter Drone

Technological systems consist of interacting components. Some new systems develop from novel configurations of existing components. The quadcopter drone might be considered as a novel system in this manner. U.S. patent 3,053,480 for the quadcopter drone is shown in Fig. 9.23. A patent diagram shows the now familiar aircraft as a unique combination of four driven propellors and an appropriate subsystem for lift, stabilization, and control. Current implementations have migrated to individual motors for each propellor rather than a centralized power source, however essential system features remain recognizable.

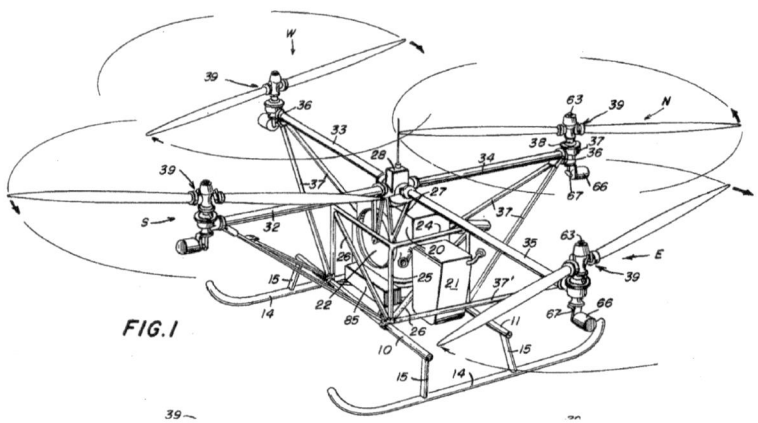

**Fig. 9.23**  U.S. Patent 3,053,480 omni-directional, vertical-lift, helicopter drone. Public domain

### Bibliography

Ackerman, Evan, "A Brief History of the Microwave Oven" *IEEE Spectrum*, 30 September, 2016. Accessed August 1, 2023. https://spectrum.ieee.org/a-brief-history-of-the-microwave-oven

Arthur, W. Brian. *The Nature of Technology: What It Is and How It Evolves.* Simon and Schuster, 2009.

Basalla, George. *The Evolution of Technology.* Cambridge University Press, 1988.

Callister Jr, William D., and David G. Rethwisch. *Materials Science and Engineering: An Introduction.* Wiley, 2018.

Crouch, Tom D. *Wings: A History Of Aviation From Kites To The Space Age.* W. W. Norton & Company, 2004.

Engel, Jerome. *Clusters of Innovation in the Age of Disruption.* Edward Elgar Publishing, 2022.

Headrick, Daniel R. *Technology: A World History.* Oxford University Press, USA, 2009.

United States Patent and Trademark Office (USPTO) "Patent Basics."Accessed August 1, 2023. https://www.uspto.gov/patents/basics.